Street-by-Street Retrofit

For many years, it has been recognised that improving the energy performance of the existing housing stock is vital if energy demand is to be reduced to combat climate change. The art of retrofit is posited as a way forward beyond today's weak pseudo-Modernist architecture – all that is left – the final echo of Modernism's original utopian impulse.

Central to the book is the presentation of domestic street-by-street retrofit as an issue with technical, financial and societal dimensions. A holistic view of the complex, interacting factors that have held back any advance is interspersed with a historical account of retrofit's faltering progress over the last 20 years. The crucial challenges that have been encountered are described, including the technological and human factors that urgently need to be addressed. It is suggested that the utopian instincts that propelled early Modernism can be redeployed in finding an approach to retrofit that will pave the way towards a politically engaged architecture of social purpose.

Street-by-Street Retrofit's goal is to involve the creative imagination of designers and form an alliance with policymakers and many others in the business of urban improvement; it is intended for all these audiences.

Mike McEvoy studied at Cambridge prior to registration as an architect, then went on a postgraduate scholarship to Cornell; his PhD is from the Bartlett. He was in practice in the United States and Canada and for a decade with Arup Associates in London. Subsequently, he was Coordinator of Technical Studies at the University of Westminster; on the faculty at Cambridge and a Fellow Commoner of Downing College; and latterly, Professor of Architecture at the University of Brighton, where he led EU IFORE an Anglo/French €6.3 million street-by-street retrofit programme (which is the background to this book). Previously, he had completed, and published, the outcomes of several funded research projects into low-energy construction. He has written three other books on architectural technology: *Architecture and Construction in Steel*, *External Components* and *Environmental Construction Handbook*.

Routledge Research in Architecture

The *Routledge Research in Architecture* series provides the reader with the latest scholarship in the field of architecture. The series publishes research from across the globe and covers areas as diverse as architectural history and theory, technology, digital architecture, structures, materials, details, design, monographs of architects, interior design and much more. By making these studies available to the worldwide academic community, the series aims to promote quality architectural research.

Between Theory and Practice in Architectural Design
Imagination and Interdisciplinarity in the Art of Building
Alexander Tsigkas

Mies Contra Le Corbusier
The Frame Inevitable
Gevork Hartoonian

Architecture in the Age of Mediatizing Technologies
Sang Lee

Eating, Building, Dwelling
About Food, Architecture and Cities
Edited by David Arredondo Garrido, Juan Calatrava and Marta Sequeira

Photography, Architecture, and the Modern Italian Landscape
Primitivism and Progress
Lindsay Harris

Street-by-Street Retrofit
A Future for Architecture
Mike McEvoy

For more information about this series, please visit: https://www.routledge.com/Routledge-Research-in-Architecture/book-series/RRARCH

Street-by-Street Retrofit
A Future for Architecture

Mike McEvoy

LONDON AND NEW YORK

Designed cover image: Erdington from above. Credit: Manuel Perez Galavis

First published 2025
by Routledge
4 Park Square, Milton Park, Abingdon, Oxon OX14 4RN

and by Routledge
605 Third Avenue, New York, NY 10158

Routledge is an imprint of the Taylor & Francis Group, an informa business

© 2025 Mike McEvoy

The right of Mike McEvoy to be identified as author of this work has been asserted in accordance with sections 77 and 78 of the Copyright, Designs and Patents Act 1988.

All rights reserved. No part of this book may be reprinted or reproduced or utilised in any form or by any electronic, mechanical, or other means, now known or hereafter invented, including photocopying and recording, or in any information storage or retrieval system, without permission in writing from the publishers.

Trademark notice: Product or corporate names may be trademarks or registered trademarks, and are used only for identification and explanation without intent to infringe.

British Library Cataloguing-in-Publication Data
A catalogue record for this book is available from the British Library

Library of Congress Cataloging-in-Publication Data
Names: McEvoy, Michael, author.
Title: Street by street retrofit : a future for architecture / Mike McEvoy.
Description: Abingdon, Oxon ; New York, NY : Routledge, 2025. | Series: Routledge research in architecture | Includes bibliographical references and index.
Identifiers: LCCN 2024035986 (print) | LCCN 2024035987 (ebook) | ISBN 9781032932262 (hardback) | ISBN 9781032949482 (paperback) | ISBN 9781003564997 (ebook)
Subjects: LCSH: Architecture and society--Great Britain. | Architecture and climate--Great Britain. | Buildings--Retrofitting--Great Britain.
Classification: LCC NA2543.S6 M387 2025 (print) | LCC NA2543.S6 (ebook) | DDC 720.1/030941--dc23/eng/20240923
LC record available at https://lccn.loc.gov/2024035986
LC ebook record available at https://lccn.loc.gov/2024035987

ISBN: 978-1-032-93226-2 (hbk)
ISBN: 978-1-032-94948-2 (pbk)
ISBN: 978-1-003-56499-7 (ebk)

DOI: 10.4324/9781003564997

Typeset in Times New Roman
by KnowledgeWorks Global Ltd.

Contents

List of Figures   *viii*
Foreword: What If the Just Transition Began at Home?   *ix*
Acknowledgements   *xi*

Introduction   1

PART I
The State of the Art   3

1 Which Way to Jump?   5

Case Study: Energiesprong 9

2 The Carrying Capacity of the Planet   12

The Anthropocene and nature 12
Sustainable development, its origins and implications 13
Ecomodernism and the magic of technology 13
The climate crisis and the crisis of culture 14
Cultural juncture 14
Case Study: EU IFORE Innovation for Renewal 15

3 Re-evaluation of Modernism   19

Tipping point 20
Regulation 20
Alternative directions 21
The ecomodernist direction: Geo-engineering 22
Case Study: Parity Projects 23

4 Sustainable Retreat 26

Technology (alone) is not the answer 27
The future slowdown 27
The idea of progress 28
Modernism and the natural world 29
Architecture as a belief system 30
Case Study: Link Road, Birmingham 31

5 Caring Architecture 35

The future role of the architect will be to build sparingly 36
Building little implies making good what we already
  have i.e. retrofit 37

PART II
A Brief History of Retrofit 39

6 Size of the Problem 41

The energy case for retrofit 43
Problem #1 Determining the Outcomes 44

7 Pre-requisites for Retrofit 60

Retrofit at the urban scale 60
Retrofit origins 61
The 40 per cent house 61
Problem #2 Insulation 62

8 Drivers for Change 69

Problem #3 Overheating 70

9 Home Truths 76

40 per cent house to achieving zero 77
Problem #4 Ventilation 78

10 Anticipating the Green Deal 86

Problem #5 Airtightness 89

11 Retrofit Comes to a Halt 96

Problem #6 Renewables 97

12 Measuring Success: 80% Reduction and 'Retrofit for the Future'   107

Low carbon Britain 110
Problem #7 The occupants 111

13 The Progress of Retrofit   123

Retrofit and performance 124
Architects as retrofit leaders 126
Architectural skills required? 126

PART III
Towards a New Utopia   129

14 The Art of the Imagination   131

Retrofit realigned 133
Retrofit and utopia 134

15 The Problem of Theory   138

The politics of architecture 141
Capitalism and creative destruction 142
Modernism and radical politics 143

16 Architecture's Very Uniquely Compromised Position   144

*The roots of Modernism: Hannes Meyer and the New
   Objectivity (Neue Sachlichkeit) 146*
*The hidden aspects of consciousness, the uncanny,
   Gothic and Surrealism 149*
Urban ecology 151

17 Echoes from the Past: Herbert Marcuse   158

Society as a work of art 159
Retrofit as the representation of society as a work of art 160
Marcuse's utopia of hope, utopia as a realisable dream 161
Retrofit as subversive art 162

18 Retrofit and Architects: A Future   164

Architects and innovation – Our utopian mission 165
Architecture or extinction 168

Index   *169*

List of Figures

16.1	One of Ernst Bloch's 'hollow spaces of capitalism'. Credit: the author	146
16.2	Hannes Meyer: Entry to the League of Nations Competition, 1926–1927. Credit: Reproduced from Claude Schnaidt, Hannes Meyer: Buildings, Projects and Writings (Teufen: Verlag Arthur Niggli, 1965)	148
16.3	André Breton in front of the painting by Giorgio de Chirico, "L'énigme d'une journée", 1922. © MAN RAY 2015 TRUST/ADAGP-ARS – 2024, image: Telimage, Paris. Copyright: For Man Ray – © **Man Ray 2015 Trust/DACS, London 2024** For de Chirico – © **DACS 2024**	151
16.4	Carved capital in the form of hawthorn leaves, from the Chapter House at Southwell Minster, c. 1330. Credit: By Andrewrabbott – Own work, CC BY-SA 4.0, https://commons.wikimedia.org/w/index.php?curid=37918852	153
16.5	Four Gables, Brampton, Cumberland, c. 1878. Philip Webb's reinterpretation of a border country fortified tower house. Credit: the author	155
17.1	Herbert Marcuse addresses the Angela Davis Congress in Frankfurt, June 6, 1972. Credit: Keystone Press/Alamy stock photo	158

Foreword: *What If the Just Transition Began at Home?*

We are alive at a moment of flux, of cascading and overlapping crises – from the local to the global – of crumbling colonial facades and possibility. The climate and ecological emergency shapes billions of lives without reprieve – disproportionately harming vulnerable communities and the countries shouldering least responsibility for its acceleration,[1] and the consequences of the UK's deepening (political) culture of uncare for its built environment and those living within it has reached a fever pitch.[2] Costs of energy,[3] renting[4] and survival[5] continue to be raised to increase profits,[6] and after decades of insufficient, minimal, or reactive maintenance, home is not guaranteed to be a place of comfort, health, nor safety,[7] and hundreds of thousands of people across the United Kingdom cannot afford access to housing.[8] But the good news is, this is not inevitable: We have sufficient technology, experience and creativity to go far beyond painting over these interconnected positive feedback loops of late-stage capitalism. If we were to co-design and steward the climate transition with care and organise collectively around street-by-street retrofit endeavours, we could radically transform life for millions of people.

The retrofit of the millions of homes across the United Kingdom requiring adaptations and upgrades could save many lives, reduce poor housing's pressures on the National Health Service and bolster public health, lower energy demand locally and nationally, reduce fossil fuel dependencies, and would require the creation of training programmes, supply chains, infrastructures and many, many jobs.[9] Properly resourcing and investing in a task of such scale could feasibly lay the foundation for a Just Transition that leaves nobody behind. Should we expand our retrofitting imaginary to consider streets as living systems and something that neighbourhoods do together,[10] then even more becomes possible, and things get even more exciting. Should we approach it compassionately and imaginatively, the scale of the positive impacts that the adaptation of our communities and housing to our rapidly warming world, and the retrofit of our homes, streets and neighbourhoods, could generate are dreamlike.

As we collectively rise to that challenge, architectural skillsets, experience and education could contribute, even more, if reimagined in service to, and in the spirit of that Just Transition. Architecture's practitioners, educators and learners can offer communities visualisations of possible futures, embody roles of stewardship for buildings, support those doing grassroots community organising, demystify the languages of policy, systems and technology, and more.

To harness the full transformational potential of street-by-street retrofitting, we need to make time to learn from examples – successes and failures – through books such as this, and we must work alongside those who have experienced fuel poverty or displacement, resisted demolition, eviction or gentrification, campaigned for better housing and working conditions, and stood up for tenants' rights. Let us begin with, and return often to, imagining what possible futures could feel like; first asking ourselves, and then those around us, the core question of Retrofit Reimagined: *What if the climate transition and retrofit of our homes and streets were designed, owned and governed by the people who live there?*

Scott McAulay is the founder and coordinator of the Anthropocene Architecture School and works as a Regenerative Design and Infrastructures Specialist with Architype.

Notes

1. Jason Hickel. "Quantifying National Responsibility for Climate Breakdown: An Equality-Based Attribution Approach for Carbon Dioxide Emissions in Excess of the Planetary Boundary." *The Lancet*, 2020. www.thelancet.com/journals/lanplh/article/PIIS2542-5196(20)30196-0/fulltext.
2. Peter Apps. Show Me the Bodies: How We Let Grenfell Happen. London: Oneworld, 2022. p. 310.
3. CIVIC SQUARE and Dark Matter Labs. "3°C Neighbourhood." 2024. https://drive.google.com/file/d/11JLLVqvHeh1c4FULWTJqwjlc0qQ031XO/view.
4. Office for National Statistics. "Index of Private Housing Rental Prices, UK: January 2024." *ONS*, 2024. https://www.ons.gov.uk/economy/inflationandpriceindices/bulletins/indexofprivatehousingrentalprices/january2024.
5. The House of Commons Library. "Rising Cost of Living in the UK." 2024. https://researchbriefings.files.parliament.uk/documents/CBP-9428/CBP-9428.pdf.
6. End Fuel Poverty Coalition. "Energy Profits Hit £420 bn in Recent Years as Standing Charges Rise." 2024. https://www.endfuelpoverty.org.uk/energy-profits-hit-420bn-in-recent-years-as-standing-charges-rise/.
7. British Research Establishment. "The Cost of Poor Housing in England by Tenure." 2023. https://bregroup.com/documents/d/bre-group/bre_cost-of-poor-housing-tenure-analysis-2023-pdf#:~:text=The%202021%20research%20found%20that,%C2%A318.5%20billion%20per%20annum.
8. The Big Issue. "Homelessness Facts and Statistics: The Numbers You Need to Know in 2024." 2024. https://www.bigissue.com/news/housing/britains-homelessness-shame-cold-hard-facts/.
9. Passivhaus Trust. "The Retrofit Challenge." 2020. https://www.passivhaustrust.org.uk/competitions_and_campaigns/passivhaus-retrofit/#Passivhaus%20Retrofit.
10. CIVIC SQUARE and Dark Matter Labs. "The Great British Energy Swindle." 2022. https://provocations.darkmatterlabs.org/the-great-british-energy-swindle-92a059a49248.

Acknowledgements

With particular thanks to Drs Arianna Sdei and Ryan Southall, and to my co-workers Sam Lawton, Elisabeth Lécroart, Sue Robertson and Professor Susannah Hagan; to Adam Nicolson for his insights and advice, and to Tana as always.

Acknowledgments

Introduction

It was in 1964 that Sir Leslie Martin, architect of the Royal Festival Hall and then current Professor of Architecture at Cambridge, was asked by the Labour government to draw up plans for the demolition and rebuilding of Whitehall.[1] Architecture stood side-by-side with Harold Wilson in his wish to remake society in the 'white heat of technology.' Much would have been avoided, the MPs' expenses scandal for one, each member of parliament would have had an in-town flat provided, and incidental embarrassments such as the colonial-era murals adorning the Foreign Office would have been swept away. But architects are useful for tilting at windmills, and this scheme proved to be one. The last half century has seen architects disappear from the high table. Each decade has contributed to the swan song, the decline of public housing, the undermining of expertise, a reactionary upper class, the moral bankruptcy of design and build, and most recently the horror of Grenfell Tower. Over the same time period, though, adherence to the notion of Green Architecture has grown, so there are now bookshelves of available titles. The great conundrum is why there is nothing recognisable as 'Green Architecture,' architecture is after all a visual art? True, there are any number of apartment buildings clad with shoddy timber cladding, but where is it to be found amongst the high culture of architecture. Most certainly nowhere within the many books where the word 'eco' is affixed to resorts, hotels and expansive private houses surrounded by lush green vegetation, or technical guides filled with anxious charts, diagrams and wagging fingers.

The pretence has been throughout Modernism, from Baudelaire onwards, that the arts are hermetic disciplines removed from political debate. But, Modernism was energised by the upheavals following World War I, particularly in those countries undergoing revolution; political discord was rife at the Bauhaus. In the UK, where the old order prevailed, it wasn't until the end of the following war and the resulting establishment of the welfare state that Modernism was adopted as the clean and logical solution to the built representation of socialism. Once, however, the United States became the leader in progressive design, following the emigrant first-generation masters fleeing from Europe, architecture was soon shorn of its social purpose. Johnson and Hitchcock's book *The International Style* clipped any

DOI: 10.4324/9781003564997-1

left-wing associations and later, under the influence of Venturi, the discipline became entirely self-referential if temporarily re-energised by casting around for new sources and inspirations. Pseudo-Modernism the style that has become universal in recent years, further emaciated by minimalism, is the latest (and last?) recycling of a tired and increasingly thin formal language now just the wrapping for real estate and monuments to corporate greed.

For years it has seemed as if we are waiting for that point of change, the impetus that will kick-start the next chapter of the story of architecture. Meanwhile, the world around us is subject to a series of convulsions. It is only too obvious, particularly to the young, that shibboleths such as 'Sustainable Development' are not fit for purpose. The rest of the world is anxious to raise living standards towards that of the West's middle class, even though billions still live in poverty, such that the capacity of the world to cope has already been exceeded. Capitalism's debt-driven need for unfettered growth and development has been recognised by Extinction Rebellion as outmoded and unsustainable. But this is not just a matter of economics, the cultural foundations of the Western world are being shaken, questions of belief that are inextricable from political considerations.

For architecture, the consequences are profound; if the West now needs not just to operate differently but to offer an innovative, responsible yet enticing paradigm for wider adoption, how might this be represented in the form of buildings and cities? After architecture's history – celebrating power, colonialism and conquest – then after the globe has been spread with rampant materialism and individualism (and their accompanying architecture), perhaps there is now a chance for architects to re-imagine architecture by discovering urbanity as they find it and to begin working towards its improvement, socially, visually and environmentally, through street-by-street retrofit.

Note

1 The scheme's significance is the subject of Sharr, A and S. Thornton. 2016. *Demolishing Whitehall: Leslie Martin, Harold Wilson and the Architecture of White Heat.* London: Routledge.

Part I
The State of the Art

1 Which Way to Jump?

> One of the greatest contrasts between our times and the period between the two Great Wars is that they had a sense that progress could be anticipated, and that the Zeitgeist could be made manifest. Architects in the 1920s and 1930s were able to think in utopian terms about the city because they felt themselves capable of giving shape to the spirit of the times. That is something that we don't dare to do today, when only the most radical pragmatism seems to prevail. Rafael Moneo: Architecture Culture Needs a New Paradigm, Soane Annual Lecture 2017.
>
> (Moneo 2017)

Rafael Moneo expresses a commonly held view, idealism is dead, the role of the architect is to roll over and give speculators what they want. Any number of books have been written in recent years bemoaning architecture's current situation. Miles Glendinning in *Architecture's Evil Empire: The Triumph and Tragedy of Global Modernism* (2010) points to the continuing rash of energy-guzzling glass skyscrapers and asks when the green agenda will start to make them obsolete. He maps the sad story from the original social goals of Modernism to the current parade of aggressively individual icons, each one extravagantly 'special' yet each depressingly the same, contributing to the worldwide homogeneity of the global city.

Similarly, Tom Dyckhoff in *The Age of Spectacle* (2018, 351) portrays current architecture, recycling Modernist aesthetics, but now in a new guise, reduced to all but its easily tolerated aspects. Architects have become apologists for, and enablers of, the entrepreneurial city of spectacle:

> The age of spectacle – and the free market that drives it – is still with us. Though the free market in the gentrified city is hardly free, just open to the highest bidder. What power do any of us have to say yes or no to this building or that? The right to the city has been replaced by the right to buy it … The free market anticipates, stimulates, manipulates and delivers the desires of its citizens, for a price. It just lacks the freedom bit … Life in the modern city remains a utilitarian drudge, a highly efficient machine for squeezing work and money out of their inhabitants.

DOI:10.4324/9781003564997-3

In *Four Walls and a Roof* (2019, 417–8), from the front line of practice, Reinier de Graff lays bare the current predicament of architects. Comparison is made between the path followed by architecture and that of *Capital in the 21st Century*, the now celebrated analysis by Thomas Picketty of the returns on capital, unearned income, with those of paid employment through the last 100 years. For a period, from between the wars to the 1980s, wages increased as did social mobility, the returns on property were relatively modest and the class divide shrank. During this time, when unions were strong, and communism was on offer as a competing alternative, the goodwill of workers was maintained by the regulation of markets and a generous welfare state. Modernism's heyday coincided with this period: A shared belief in progress, faith in the benefit of technology as a public good, and the relentless pursuit of the 'new' for the betterment of all. Perhaps though, that 50-year interval was just an anomaly in the history of capitalism; up to World War I, returns on capital outdid those of labour; the chance of growing rich through work was outstripped by returns on wealth, so social mobility was that much harder. This is the situation we have returned to since the disintegration of the Soviet bloc, and the neoconservatism ushered in by Reagan and Thatcher that became the neoliberalism of today. Architecture has followed down the same path. Whereas the stripped-down aesthetic of Modernism was intended to make dwelling affordable, public housing projects were the mainstay of architecture and the background fabric of the modern city, now the market has taken control. Our contemporary pseudo-Modernism is just cheap construction geared to maximising the profits on development. Property, rather than providing a place to live, has become an asset class; architecture is now assessed by its market value. Superficially, the two Modernisms, heroic and pseudo, bear some similarity, but their underlying values are completely different. The idealism of the early moderns has been replaced by value-engineering and the relentless urge for profit. The story of architecture has now become the story of economics and capital.

> 'There is a view that capitalism has failed – and that it needs somehow to be totally transformed. My desk is littered with books explaining the miseries of capitalism; some offering radical solutions for saving or reimagining it. It's mostly unnecessary nonsense.' Merryn Somerset Webb author of *Share Power: How ordinary people can change the way capitalism works – and make money too*, quoted in the Financial Times 15th January 2022.

Whilst the trajectories of economics and architecture have paralleled one another since the start of our current era, the age of neoliberalism, in both cases diagnoses and prescriptions reflect the spectrum of political opinion. Merryn Somerset Webb presents the long-established, establishment view that summarises as: Following the end of the Roman Empire, 1000 years of short and brutal lives were transformed by the Enlightenment, since when science and capitalism have enabled longer life and rescued millions from poverty. The market mechanism is constantly rising to new challenges, and given past performance, it will undoubtedly find its way to a better future, particularly now that it is fully attuned to the climate challenge. This mindset is exemplified by the so-called Skidmore Report from 2022 that defined the UK government's approach to the environmental challenge as being both pro-business and pro-growth.

Set against this is the welter of books on Merryn's desk, often by economists retreating from the failings of the 'dismal science' of economics. New models are proposed that seek to establish new measures of human happiness and well-being to replace the previous belief in the 'rational' pursuit of financial self-interest. Not only has wealth been shown to not equate with happiness, or at least once beyond that of providing a secure existence, but also current levels of inequality are corrosive to the fabric of society. The relentless quest for growth has only been achieved at the expense of the planet; environmental costs aren't factored into profit and loss accounts. Airline companies don't pay for the air they pollute, nor do cruise liners pay for their use of the sea.

> A member of the team that published *Limits to Growth*, Jorgen Randers, recently offered a view of the near future in a book entitled 2052: A Global Forecast for the Next Forty Years. In it, he predicts a continuation of the global divide between rich and poor, with a minority securing affluent lives at the expense of the rest. Our current global system, he believes, promising prosperity through continued economic growth, offers false hope to most of the human race. 'To this day,' he warns, 'six billion people are being misled into believing that there are no natural constraints and they can have it all because human ingenuity will come to the rescue. The truth is they simply cannot.'
>
> (Lent 2017, 429)

Two recent examples of the books dismissed by Somerset Webb are *Tomorrow's Economy* by Per Espen Stoknes (2021) and *Climate Leviathan* by Joel Wainwright and Geoff Mann (2020). Stoknes, a colleague of Jorgen Randers at the Norwegian Business School, is equally sceptical about current capitalism – based on relentless growth in GDP, the route to ecological collapse; also 'de-growth' that would result in unpayable debt, growing unemployment and inequality, without the means to finance innovation; and 'sustainable development,' somehow possible even whilst non-renewable resources are being exhausted, a term now so devalued as to now just be a catch-all excuse for greenwash. Stoknes would rather have a form of growth more akin to the way trees grow and exchange resources with the environment for their mutual benefit. Stoknes investigates four scenarios from our current 'grey growth' based on planetary exploitation to his preferred 'green growth' with GDP still increasing but accompanied by diminishing use of resources. He proposes a 'smarter' solution to attaining sustainable development goals through 'green growth,' requiring most governments around the world to implement five transformative measures: The total adoption of renewable energy (but he doesn't mention the environmental cost, for example, the 500,000 gallons of water used for the extraction of one tonne of lithium) reconstructed food chains based on plant-based diets, new tree-like patterns of development involving the smart use of resources and a circular economy, active inequality reduction, and increased investment in education. Stoknes concedes that no country at present is performing in line with his goals.

Wainwright and Mann also investigate four possible futures, but theirs is the world of politics, within which they find many risks. The first of the four, they maintain the most likely contender, is the consolidation of power by the establishment class of the liberal democracies, perhaps in collaboration with China within a transnational entity such as the UN, to achieve global control of 'green capitalism' using aggressive methods of surveillance and discipline. It's ethos 'faith in progress.' The second, in which the core capitalist states succumb to right-wing populism resulting in either climate denial or unwillingness to regulate or intervene, in the cause of so-called 'freedom,' involving integration into the security process of terror and repression. The third, authoritarian state socialism evoking 'confidence in the masses,' a command economy brought about in the first instance by revolution in Asia, where the world's largest populations are also those subject to the worst effects of climate change. Finally, their admittedly utopian alternative represents a socialist outcome, committed to a communitarian, egalitarian future and emerging from the overwhelming protest on the streets as a result of activism and agitation.

> Sustainable development is a crazy notion. It was a wonderful idea in 1800, when there were only a billion of us on the planet, and if we'd adopted it then we wouldn't be in the mess we're in now. Now it's far too late.
> (Lovelock 2011, 35)

So, within this challenging context, it is high time that the notion of 'Sustainable Architecture' was reassessed. It came out of the Brundtland Commission's coining of the phrase Sustainable Development, synonymous with the business-as-usual mirage of capitalism's 'grey growth.' Our current model of capitalism the planet can't afford now, let alone as the rest of the world struggles to achieve the living standards of the West. After many years when operational energy was perceived to be the problem, now embodied energy has come into focus, as if attaining zero carbon construction was the way to a sustainable future. Meanwhile, the triple crisis of climate change, biodiversity loss and pollution gets worse. Global Footprint Network estimates that 80 per cent of the world's population already lives at a level the planet can't support and that by 2050 an expected population of 9.7 billion would require the equivalent of three planets. The realisation that sustainability's attempts at amelioration have been of limited impact has led to the call for 'regenerative' thinking. As espoused in *Sapiens* (Harari 2015), humanity's problems date back not from the Industrial Revolution but from 10,000 years ago at the agricultural revolution and the first cities. Regenerative thinkers would like a fundamental realignment to achieve the symbiosis with nature that is still achieved by the world's few remaining hunter-gatherers, most certainly a utopian project.

> The housing question that still haunts architecture and development on a global scale; the question of density raised by population explosions and land scarcity; the ecological crisis in resources and modes of conservation that,

with radical shifts in climate and diminishing energy sources, presents more fundamental problems for architecture than those addressed by development in materials and 'green building' alone.

(Vidler 2008, 153)

Given the threats of climate, social and ecological catastrophe, clearly from here-on-in all fossil fuel should be left in the ground, and the job of the Western architect should be to build as little as possible. Retrofit, making the best of what we have, is slowly moving centre stage but has yet to become architecture's main line of business. The middle class of the West has achieved material comfort, so our buildings should only be replaced under extreme circumstances as determined by democratic process. Design for retrofit, rather than being about the unfettered tabula rasa of Modernism, is a matter of caring and improving, engaging with communities for the enhancement of their lives.

Looking back through human history, we can recognize two great phase transitions: the emergence of agriculture about 10,000 years ago and the rise of the scientific age over the past few hundred years. In our current era, it's becoming clear that we're entering a third great transition.

(Lent 2017, 408)

Case Study: Energiesprong

Pioneered by the Dutch government, Energiesprong (energy leap) (Energiesprong n.d.) is an over-cladding system that encapsulates the walls and roof of existing dwellings with dry assembly panels. The new enclosure is both insulated and airtight, incorporates solar PV on the roof and an 'energy hub' with either an air or ground source heat pump and optional batteries. The panels are made using dimensions from a laser survey, which forms the basis of a BIM model for the precise manufacture of insulated units with an exact fit. An early application in England was the retrofit of a terrace at Thamesmead, between 2013 and 2015. The prefabricated wood-framed cassettes, made by a Bavarian timber construction company, were filled with 28 centimetres of cellulose insulation and came to the site with integral high-performance windows. The panels were craned into position and bolted off the existing structure whilst the occupants remained in residence. A pitched roof was built over the existing flat one, bringing the whole building up to EnerPHit standard.

A similar scheme, a four-storey terrace of maisonettes over garages, was completed in Coventry. Communal heating and solar thermal were considered, but because individual bills were a requirement, instead, there is a boiler in each dwelling, and heat recovery ventilation was installed throughout. Not surprisingly, these first demonstration projects were expensive, but the community focus of the projects in terms of training and employment added value, such that the German company intended to start making the cassettes in the United Kingdom and to establish efficient partnerships with local contractors (UKGBC n.d.).

That was back in 2014, but since then progress has been slow. Energiesprong UK's feedback on completed projects reported that the industry remains underdeveloped, with components not always being correctly specified or properly installed. Also, the long-standing problems with monitoring systems that don't give a consistent output for adequate comparison between sites. This is complicated by the wide range of results recorded because of the different lifestyles and varying degrees of engagement between households, including varied preferences in terms of indoor temperature. And the perpetual problem remains that although rocketing heating costs have put the need for energy efficiency into focus, the means to pay for it continues to be elusive.

More recently, in 2017, the system was used for a pilot project in Nottingham retrofitting seven houses and three bungalows, including the trialling of different service solutions; the retrofits took only a week with the occupants remaining in their homes. This success has resulted in Energiesprong being adopted as the strategy for further schemes in Nottingham, so now 60 units have been completed and 150 more are in the offing. A local factory has been established to produce the panels using Modern Methods of Construction (MMC) methods.

Another demonstration project, this one in Devon, was completed in 2020 and included the retrofitting of 15 homes. Part of the rationale for the local authority was that the majority of 'low hanging fruit,' loft and cavity wall insulation had now been completed under CERT and CESP and currently ECO. But over the course of time, tackling these measures in more difficult cases has trebled the costs per tonne of carbon saved when comparing ECO with the earlier funding schemes. The case made for Energiesprong being that 'whole house' retrofit is a prerequisite for the United Kingdom to achieve its carbon reduction targets. Scaled up to the needs of the United Kingdom, compared with the results from conventional approaches applied to 15 million homes, Energiesprong applied to only 11 million would triple the achievable carbon reductions. This level of intervention could lessen national heat demand by over 40 per cent, significantly reducing the cost of decarbonising the grid. In Devon, the predicted outcome was for an increase in property values of 25 per cent, and a reduction of over 60 per cent in energy bills (Green Alliance 2020).

Energiesprong has formidable advantages in achieving 'deep retrofit' upgrades in one operation, as now regarded as a necessity by London Energy

Transition Initiative (LETI) for example, without having to decant the residents, a major consideration. The panelised, timber-based panels have additional advantages in demountability and end-of-life recycling. At present, the system is expensive, but costs in the Netherlands have reduced over time; factories have achieved labour productivity improvement of 75 per cent and waste reduction of 90 per cent since inception in 1994. By comparison, the productivity of UK on-site construction scarcely changed over the same period. Large-scale adoption would undoubtedly bring down costs as it has in the case of wind power and solar PV. It is the first major introduction of off-site production methods, some years after The Energy Zone Consortium, a since disbanded group of energy companies, housing associations, architects, academics and construction companies, investigated the potential for Modern Methods of Construction (MMC) in retrofit. The fragile linkages between customer information and analysis, inventory, factory processes, quality assurance, monitoring and control will require government commitment. Market pull and regulatory push are needed to steer it through the boom-and-bust cycles that have bedevilled other attempts to direct construction away from on-site wet trades.

References

de Graaf, R. 2019. *Four Walls and a Roof: The Complex Nature of a Simple Profession*. Cambridge, MA: Harvard University Press.
Dyckhoff, T. 2018. *The Age of Spectacle: The Rise and Fall of Iconic Architecture*. London: Windmill Books.
Energiesprong. n.d. Available online: https://www.energiesprong.uk (accessed 7 June 2022).
Glendinning, M. 2010. *Architecture's Evil Empire: The Triumph and Tragedy of Global Modernism*. London: Reaktion Books.
Green Alliance. 2020. *Reinventing Retrofit: How to Scale up Home Energy Efficiency in the UK*. Available online: https://green-alliance.org.uk/wp-content/uploads/2021/11/reinventing_retrofit.pdf (accessed 7 June 2022).
Harari, Y.N. 2015. *Sapiens: A Brief History of Humankind*. NSW Australia: Generic.
Lent, J. 2017. *The Patterning Instinct: A Cultural History of Humanity's Search for Meaning*. New York, NY: Prometheus.
Lovelock, J. 2011. '*Looking into the Future*' Quoted in the *RIBA Journal*, April:35.
Moneo, R. 2017. *Soane Medal Lecture 2017*. London: Sir John Soane's Museum.
Stoknes, P.E. 2021. *Tomorrow's Economy, A Guide to Creating Healthy Green Growth*. Cambridge, MA: MIT Press.
UKGBC. n.d. UK Green Building Council: Energiesprong. Available online: https://www.ukgbc.org/solutions/energiesprong-uk (accessed 10 June 2022).
Vidler, A. 2008. *Architecture's Expanded Field in Architecture between Spectacle and Use*. Clark Studies in the Visual Arts. New Haven, CT: Yale University Press.
Wainwright, J., and G. Mann. 2020. *Climate Leviathan: A Political Theory of Our Planetary Future*. New York, NY: Verso.

2 The Carrying Capacity of the Planet

Opinions vary as to the carrying capacity of the planet or its likely future population (Cumming 2016), but there seems to be a consensus that if the present eight billion of us lived a Western lifestyle, the world would be too small. In China, which has become the number one producer of green technologies and is hastening their installation, renewable capabilities are far exceeded by the growing rate of consumption. The laudable aim to reduce pollution, and our impacts on the natural world, run in contradiction to policies to reduce poverty. China has achieved a sizeable middle class, whilst 800 million still live an agrarian lifestyle. Once urbanised, their consumption will be limited, but only so long as they are poor. Whereas cities in low-income nations produce few emissions per household, in high-income countries the CO_2-equivalent per person can be between six and thirty times higher. Most of the projected population growth is predicted to be in what at the moment are low- or middle-income countries.

The desire to live as does the privileged West is to achieve a better life for one's children, it's an impulse that is too strong to resist, throughout the rest of Asia and the other impoverished billions of the so-called 'developing' world. Clearly, the wealthy countries have a responsibility to adopt simpler lifestyles, a cultural move away from the accumulation of material wealth as measured by GDP towards a broader measure of human needs.

The Anthropocene and nature

It isn't necessary to revisit the extent of the problem because evidence of the changing climate is all around us, as is the result of human incursions into forests and wildlife. The urgency of the issue is illustrated by the rate of growth of carbon emissions so far this century and alarming revisions to climate models that illustrate the inadequacy of current responses. The world is currently on track to hit 2.5 degrees of warming by 2100.

So, we are now clearly living within the Anthropocene. For years, green architecture has generated bookshelves of publications about sustainability, using more timber, less energy consumption and solar panels, but now the realisation is more profound – that we have to live within our planet's means, that we have to find a way to be part of the world's ecosystem. Rather than being outside nature,

DOI: 10.4324/9781003564997-4

our responsibility within the Anthropocene is to find a new accommodation with ecological processes so humanity can revert to coexistence with the natural world.

> There is much that is amiss in modernity, and much to reject in Modernism; but we can hope only to temper the modern age and its ills, not to abolish them, as some of the green theorists suppose.
>
> (Gray 1993, 310)

Sustainable development, its origins and implications

We have an economic system driven by debt and the growth necessary to service the debt. Growth creates wealth, lifts families out of subsistence and into the middle class. The ongoing, mainstream agenda insists that 'progress' and growth will continue, increasing the planet's capacity after the introduction of better methods and less waste, following the discovery and introduction of as-yet unproven technologies.

The preconditions to achieving both aspects of the original 1987 definition of sustainability, a good life for all whilst preserving the environment, are outlined by Hannah Ritchie (Ritchie 2024). In terms of human health and welfare, great strides have been made; the proposition is put forward that with 'Green Growth' this progress can continue, both increasing wealth and restoring ecological balance. Yet the preconditions are matters of policy, paying countries to avoid deforestation, waste and fisheries management, and behaviours, changing towards plant-based diets. Whether all this can be achieved, in time, within current political and economic structures remains a question (Monbiot and Hutchison 2024).

> One trajectory with Moore's Law underwriting endless technological wizardry, promises a future in which humanity itself might be redefined. At the rate with which we are using up the earth's resources, another trajectory threatens potential disaster ahead for humanity. Which of these is more likely to be humanity's actual path? And, even more crucially, is there a way our civilisation can somehow be steered to achieve a sustainable path for humanity's future?
>
> (Lent 2017, 401)

Ecomodernism and the magic of technology

There is a school of thought that regards this as an engineering problem. According to Stewart Brand, for example, engineers, the problem solvers, will see a way through (Brand 2010). He maintains that eventually the more draconian technical solutions, seeding the upper atmosphere or the oceans, or mirrors in space, will have to be tried despite their unknown implications for the world as a whole and the impossibility of achieving global consensus. The world's large corporations will seize these opportunities. We are so late to address questions that were first posed 50 years ago, but now we are seeing the ramping up of green rhetoric. One issue that has been extensively aired is aviation. Every new plane that is constructed implies many years of kerosene-consuming pollution. But how long will it be before this argument is applied to construction? Every new building inevitably relies to a degree on that most polluting of materials, concrete, then has a lifetime of operating consumption ahead. Even worse when buildings are replaced with market-driven rapidity such as those in the City of London. So, when an embargo is placed on new construction, where will that leave architecture?

Addressing these problems has been left so late that technological development, as yet blue-sky, for example, the sudden rush to develop carbon capture and storage, will of itself be insufficient. New technologies are essential, but architecture is concerned with more than firmitas within the Vitruvian Triad, as Aalto put it: 'Architecture has an ulterior motive. Every building is intended to show that we wish to build paradise on Earth for man.' In the panic to save the world, architects may rediscover their true mission by retrofitting the world that already exists.

The climate crisis and the crisis of culture

> What is ultimately required is a shift toward a new way of finding meaning from our existence. Many visionaries and deep thinkers today recognize the need for a new global consciousness, based on an underlying and all-infusing sense of connectedness. The meaning we derive from our existence must arise from our connectedness if we are to succeed in sustaining our civilisation into the distant future: connectedness within ourselves, to other humans, and to the entire natural world.
>
> (Lent 2017, 440)

Clearly the climate crisis is a crisis of culture as well as politics, economics and ecology. In the same way that the relationship of our species to the rest of the natural world is called into question, for architects the built environment ceases to be a canvas onto which we make our mark but instead becomes the focus of nurture and care. To make better use of what we have will be the essence of this concern. Surprisingly, architects have cared little for retrofit, not meaning the artful remodelling of the industrial past into art galleries or opera venues, but rather the mass of housing that is the real challenge and the beckoning prospect.

Cultural juncture

As at any time when the tide of culture is changing, so does the mindset. Periods of excess are followed by a return to order. The frivolity of Rococo was outmoded by the French Revolution and the seriousness of Neoclassicism. Serious times require serious architecture. Frank Gehry's architecture of excess can be expected to be superseded by an architecture with radical intentions.

'Make do and mend' implies that the architects' principal concern will be with the built environment as existing, that architecture becomes a process of care and enrichment. The city becomes a cause for concern and not merely a building site. It has become commonplace to question the male agenda that has forever governed the world, which can no longer be regarded as a clean slate for unfettered assertion. For cities to be the focus of care, there is a need for a more equivocal, communal and co-operative female approach. The aesthetic implications are huge; the mess of life and the city can no longer be regarded in Corbusier's phrase as 'the fungus of the streets.'

Pseudo-Modernism is an international style. Although the design leaders known as starchitects have their own recognisable brand styles, they are duplicated around the major cities of the world with a studied rejection of context. In a world characterised by gross inequality, if a sustainable equilibrium is to be achieved, innovative solutions will be needed to remedy the vast separation between rich and poor. In the West, the emergence of a sharing society, utilising new technology, makes for a way of life that is more efficient not only in terms of resources but also in human labour. Why, for example, own a car that needs cleaning and insuring, when the app on your phone can call up a self-drive vehicle that looks after itself? In the West, even though we are saturated with everything we need, its grossly inequitable distribution requires re-balancing towards an approach of 'make do and mend.' The current lifecycle of city office buildings, for example, as little as 25 years, is totally unsustainable. This was recognised many years ago by the expression 'long-life loose-fit,' a worthy objective but one that couldn't be realised in the market-driven churn of neoliberalism. Recently, though, the state has had to reassert itself in pursuit of the common good, and perhaps the re-invigoration of architecture may follow.

Case Study: EU IFORE Innovation for Renewal (University of Brighton n.d.)

IFORE was an EU-sponsored project for the comparative retrofit of social housing in England and France; it ran between 2010 and 2014. In England, AmicusHorizon housing association made available 100 houses on a 1960s housing estate in Rushenden on the Isle of Sheppey, a very deprived area. In France, 100 houses of different types were retrofitted at Outreau on the outskirts of Boulogne.

When IFORE began, there was growing concern about future security of energy supply, reliant as it was on increasingly unstable parts of the world. The need for reduction in demand was also evident, given the poor state of the UK's housing, and the extent to which it is responsible for overall carbon emissions. Better-performing housing implies less demand, so fewer new power stations will be necessary in future. These imperatives haven't diminished in the interim. For social housing tenants, retrofit means better health, if they are no longer living in cold and draughty homes, lower costs, and consequently reduced levels of fuel poverty. The boost given to the local economy by a large-scale retrofit programme improves the prospects for local employment.

The Retrofit for the Future competition had just started, with the aim of finding new technical, if expensive solutions, using individual houses as a testbed. This paralleled the introduction of Passivhaus that is similarly concerned with a technological approach to a better standard of housing. Housing associations inevitably have restricted budgets and the need to achieve the best return on their investment. So, IFORE took a different approach, by engaging with the community, and its residents, to determine the energy and carbon savings that could be achieved cheaply by behaviour change. It was one of the first large-scale retrofit projects on either side of the Channel. Different and varying estimates had been made as to the relative importance of occupants' patterns of behaviour on energy consumption in their homes, and their use of everyday services and appliances. Despite uncertainty about the extent of these factors, their importance seemed without doubt.

Different methods were tried in the two countries. At Rushenden, six centimetres of external phenolic wall insulation with polymer-bonded render was applied. In France, a novel method was used; a factory production line was set up to manufacture timber cassettes containing insulation made from shredded clothing sourced from charity shops, a reliable means of local supply for the manufacturers. The cassettes were used as external insulated cladding with an outer surface of timber boards. The solution was a simplified form of Energiesprong, only the walls were overclad, the original roofs remaining untouched. The innovative over-cladding system was prototyped at Outreau on a street of terraced bungalows, the homes of older residents that were suitable for prototyping. At both sites, in England and France, IFORE's project teams were embedded within the communities. AmicusHorizon has a community outreach group who were based in an empty house in the centre of the estate. A number of events were organised; children and their schools were encouraged to take part, as were local youth groups.

A principal member of the team in England was the 'Green Doctor,' or her equivalent in France, the 'Energy Ambassador.' They worked with households, providing advice and enthusiasm, distributing free-issued gadgets and low-energy lightbulbs, and encouraging everyone to enjoy the comfort of their retrofitted homes in an energy-conserving way. The project enjoyed a lot of public support, and houses were made freely available to the University of Brighton team for thermography and pressure testing before and after the works.

The Green Doctor in Rushenden, handed out questionnaires to the households that were updated at intervals to assess how behavioural attitudes were, or were not, changing throughout the course of the project. What stood out, and was a success for IFORE, was the residents' perception that they had themselves been influential in seeing their energy bills reduce. At the end of the project, compared with the previous year, 64 per cent were of the view that they had been able to reduce their consumption. Two-thirds thought that this was by more than 5 per cent, as a result of adjusting the temperature setting on their thermostats, for example.

Both the academic partners, the universities in England and France, had a long-standing interest in the difficult problem of ventilation. Although it is usual when improving the energy performance of dwellings, to eliminate draughts and increase their airtightness, too often the result is poor indoor air quality, condensation and mould, and potentially poor outcomes for health. The technology of 'supply air' windows had been investigated previously by the university teams in both countries. As a result, 'supply air' windows, that were tested at Outreau, are now being made at La Rochelle, having become a catalogue item of the window manufacturer Ridoret.

Both Outreau and Rushenden are deprived areas, so one of the potential benefits of the retrofit programme was an increase in economic activity, and employment, resulting from engagement with local businesses. A shortage of work had long been a drag on living standards in both places. In Rushenden, for instance, it was decided that rather than replacing the existing windows, they would be fitted with new double-glazed units and made more airtight, a local company carried out the work.

An issue for England, but not in France, is that social housing was randomly sold off under the 'right to buy' scheme. As a result, the streets at Rushenden, retrofitted by the housing association, are 'pepper-potted' with unimproved 'right to buy' properties that are now privately rented. Efforts were made to join these landlords up to the project, but they all declined. Within a terrace, this creates a problem, which is not only visual but undermines the energy efficiency of a retrofitted house when the adjoining property has such poor performance. This 'pepper-potting' has to some extent been assuaged by the housing association tenants being able to choose their house colour from a range (following the example of traditional ordinances in places like Turin) but this is just a 'sticking plaster' solution.

It's a lot easier to monitor and quantify the benefits of physical measures applied to existing dwellings than it is to be definitive about the contribution to energy conservation that can be made by behaviour change and community-based approaches. As a result, the many advantages of an approach to retrofit, centred around the residents themselves, can be overlooked. It is a common belief that changing attitudes and behaviours are options that take too long and that, given the scale of the challenge, only technological solutions are viable. But of course, there are readily cited examples, such as the ban on smoking inside public buildings that is now an accepted social norm in many European countries. The goal for IFORE was to show that a focussed programme of engagement could realise many benefits.

IFORE's sociologists investigated these changing attitudes within the communities. Twenty-five per cent of those questionnaired, compared with results from three years earlier at the outset of the project, felt happier living in Rushenden. Also, nearly 50 per cent felt able to help other, and future, residents and share the information that IFORE had given them. Twenty-one per cent thought they were now more environmentally aware and acting accordingly, and within their households, there were others who were actively saving energy.

At Rushenden, the extent of savings was variable according to different house types. The most favourable reduction was achieved by a two-storey terraced house, the type that originally had the highest energy consumption, which had been retrofitted with both solar thermal and photovoltaic panels on their roofs. A reduction of approximately 58 per cent of carbon emissions was achieved with about 0.8 being due to the houses having been retrofitted, and 0.2 as a result of the Green Doctor encouraging lower room temperatures. For this house type, although an outlier, as good a reduction in carbon emissions was achieved as was the case for 50 per cent of the Retrofit for the Future properties, but at a tenth of the budget; implying that particularly for housing associations, employing a Green Doctor is beneficial. The least favourable outcome was a group of old peoples' bungalows that managed only 35 per cent heating reduction after retrofit; they hadn't received either solar thermal or PV panels and had relatively greater hot water and electricity consumption.

References

Brand, S. (2010), *Whole Earth Discipline*, London: Atlantic Books.
Cumming, V. 2016. *How Many People Can Our Planet Really Support?* BBC Earth, 24 March. Available online: http://www.bbc.co.uk/earth/story/20160311-how-many-people-can-our-planet-really-support
Gray, J. 1993. "An Agenda for Green Conservatism." In *Gray's Anatomy*. London: Routledge.
Lent, J. 2017. *The Patterning Instinct: A Cultural History of Humanity's Search for Meaning*. New York, NY: Prometheus.
Monbiot, G., and P. Hutchison. 2024. *The Invisible Doctrine: The Secret History of Neoliberalism*. London: Allen Lane.
Ritchie, H. 2024. *Not the End of the World: How We Can Be the First Generation to Build a Sustainable Planet*. London: Chatto and Windus.
University of Brighton. n.d. Available online: http://arts.brighton.ac.uk/projects/innovation-for-renewal-ifore

3 Re-evaluation of Modernism

At times of change, the search for the new requires re-evaluation of inherited traditions. Le Corbusier's rejection of the street led him to an inversion of the form of the pre-modern city. Set against this approach – the macho clean sweep – Jane Jacobs (Jacobs 1961) presents a contrasting female view. The disorder, which characterises the traditional city, is seen by her as having qualities worth preserving through community action. Straight lines on paper are shown as having direct consequences for human lives, as became all too apparent following the clearance and rebuilding of East London post-war and the relocation of families, and fractured kinship, to the new towns. In the current era, Jacobs, rather than Corbusier, seems the more relevant precursor. The urgent need for street-by-street retrofit of the nation's housing stock is the essence of 'make do and mend.' The extreme complexity of this undertaking will require engagement with households, and the needs of each family member if the aim of reduced energy consumption is to be achieved. At the same time, new techniques, from the photogrammetric surveying of individual properties to the computer-controlled manufacture of bespoke components, and the manipulation of large databases for matching and assembly, will be added to the marshalling and management of a large and diverse workforce. The job description requires teamwork, careful skill, empathy and modesty, which are female traits, Howard Roark need not apply.[1]

The resulting aesthetic will be an engaging assortment, the problem will be to maintain the coherence of England's predominately brick-built towns, but their transformation will also be a reinvigoration. The new architectural patchwork will preclude the snapshot of perfection. Much of the work of retrofit will only be feasible socially and economically if it is spread over time, coinciding with the end of life of existing components, and their renewal, to coincide with normal maintenance cycles. Because all the modifications and replacements will have to be appropriately sourced in terms of travel distance, embodied energy and their capability for safe disposal at the end of life, the design grammar will need to allow for change, not only as technologies improve but also as occupancy, building use and the climate changes over time. The aesthetic of Modernism, deriving from the example of Classicism, has been the image of the perfect and immutable object. Totally consistent in concept and detail, the ideal architectural artwork is unchanging (and unpopulated, or at least that is the stock image presented by architectural photography). This wasn't the case for Gothic cathedrals that were works in progress over centuries.

Tipping point

> The new is the only game in town, and by the same token it loses its meaning and lustre, and instead of moving onwards we seem to be forever stuck in the automated marketplace of the monotonously novel.
> Andreas Malm referencing Fredric Jameson

Jameson's argument is summarised in Andreas Malm's *The Progress of This Storm* (2018: 2) (Jameson 1991). During the early modern movement, the natural world was still abundant, the countryside was distinct from the new houses and factories, city dwellers would return to their rural home villages. But now cities stretch ever onwards, a homogeneous landscape, and the same cultural present. Nature no longer provides a point of reference, instead we 'are stranded in the mega-city where glass surfaces mirror each other, where images and simulacra rule over night and day ... without any material substance.'

We are at a tipping point that is not only ecological, economic and political but also a juncture for architectural culture. It is only too clear that the economic system that currently limps along is not to the benefit of either people or the planet; the series of world-spanning disruptions since the beginning of the new millennium have clearly displayed the limits to growth. Change has been long-coming, and left very late, but the immediacy of global communications has enabled institutions and even human behaviour to be capable of rapid adjustment. Modern architecture was born 100 years ago out of a society similarly reeling from post-war and post-pandemic confusion. The new modern architecture of the clean sweep, and clean lines, was the product of utopian fervour. This was the big splash in the pond that soon lost its initial thrust, so it is now perceived here on the far shore as the tiny ripple that is contemporary pseudo-Modernism, an architecture without meaning, just shrink-wrapped real estate.

> ... if the climate movement and its various allies are to make any real dent ... they probably – one of the key contentions of (Naomi) Klein – (will) have to reinvigorate, recycle, reroute utopian impulses.
> (Malm 2018, 220)

Regulation

> It is becoming uncomfortably clear that consumer self-regulation and the market will not by themselves avert environmental catastrophe ... rationing of some form is inevitable. The issue is whether it will be collectively managed, or whether it will be imposed by authoritarian means when it is already too late.
> (Fisher 2012, 89)

So far, limitations on building have required better energy performance as if purely physical responses might solve the problem. Conceivably in future, regulations will become more like those that govern works on historic buildings and national monuments, that new build will be the least favoured option, recognising that embodied energy and the way we live within buildings are also the issues. Architecture will of necessity become a very different enterprise if construction eventually becomes subject to rationing. But crucially, this is a cultural question, not just the inheritance from Modernism, or the industrial age, or the Enlightenment notion of progress, but our collective psychology with regard to the world around us.

Alternative directions

> We have palaeolithic brains, we have medieval institutions and space-age technologies.
>
> EO Wilson

Architecture's problem now is knowing which direction to follow. On the one hand, ecomodernists point to the growing efficiencies within technological processes that are enabling human needs to be met with diminishing reliance on natural resources. This progressive decoupling will enable the natural world to be freed from humanity's dependence upon it. However, because improvements in living standards are closely correlated with increasing energy consumption, they assert that the only practical method of climate mitigation is the development of new energy technologies. According to this theory, technology proceeds faster than mankind's abilities to transform ways of life, human institutions and politics, the scale of the challenge necessitates that technology will take the lead.

On the other hand, the environmental movement has always hoped for a future of reduced consumption, a collective and simpler way of living. The ecomodernists counter that the genii can't be put back in the bottle, the western middle-class lifestyle that is the hope of the developing world won't be stopped, so technology has to find a way of making the planet cope with the demands of a still growing population.

> Western civilisation (more recently becoming global civilisation) has followed a path of cognitive separation. By prizing reason over emotion, splitting human existence into mind and body, and then defining humanity only by its mind, we set the cognitive foundation for the scientific and industrial revolutions that transformed the world. In our relationship to the external world, we pursued a similar path of disconnection, finding meaning in transcendence while desacralizing the earth, creating root metaphors of nature as an ENEMY TO BE CONQUERED and a MACHINE TO BE ENGINEERED. Since then, we've been busy developing technologies designed to turn those metaphors into fact.
>
> (Lent 2017, 441)

The ecomodernist direction: Geo-engineering

Ecomodernist Stewart Brand (Brand 2010) takes issue with environmentalists, and their 'romantic pessimism,' who have blocked the introduction of genetically modified foods and nuclear power, technologies he advocates as both safe and necessary. In the case of geo-engineering, he recognises that development has been limited because of the obvious risks involved in treating the whole planet as an experiment. But he maintains that geo-engineering is an inevitability because of the difficulty of dismantling the current oil and gas infrastructure and the land mass and enormous costs required to achieve equivalent energy production using renewables. As the effects of climate change become more extreme, and because geo-engineering solutions are, by comparison, relatively cheap, the pressure to implement them will, he maintains, become overwhelming. So, rather than inaction, he proposes that development of all the alternatives should be accelerated, and then choices can be made between them to find approaches that are scalable and least risky. Nonetheless, our planetary problem will eventually require planetary implementation, and the consequent danger that the butterfly effect of a 'seed' in one ocean will form a 'tsunami' in another. 'Ecological awareness is awareness of unintended consequences' (Morton 2018, 16). So the dangers would of course be enormous as would the political and legal difficulties of negotiating between the winner and loser countries. Ecomodernist faith in the transformational power of technology, a belief shared with architectural dogma from Giedion through high-tech to passivhaus, contrasts with those who suggest that the world's plight is the result of human behaviours, wherein lies the basic problem we need to address (Merz et al. 2023).

Engineering methods are of course only a short-term fix to buy some time, they do nothing to prevent the ongoing destruction of the natural environment, providing only a method by which global warming can be slowed down. Brand cites the Independent newspaper's poll of 80 climatologists in 2008: 'Over half of them said that the situation is growing so dire that we must have a geo-engineering backup plan, (whilst) 35 per cent of them said that such plans would distract people from the crucial task of reducing greenhouse gases' (Brand 2010, 290).

His final declaration is… 'Ecological balance is too important for sentiment, it requires science. The health of natural infrastructure is too compromised for passivity it requires engineering. What we call natural and what we call human are inseparable. We live one life' (Brand 2010, 302).

Such is the ecomodernist direction and its perils. But most certainly, the primitivist inclinations of some green campaigners are unlikely to be realised. To imagine a world without mobile phones, even if only 20 years after their introduction, is inconceivable, let alone cars and planes. Humanity seems unable to function without technologies that once introduced make life more comfortable and convenient.

Case Study: Parity Projects

Over the past 15 years dismal history of street-by-street retrofit being put on the back burner, Parity Projects has been one of the few constants – despite the failure of the governmental initiative, the Green Deal, and years of subsequent inaction, the myriad number of reports that have been written to little effect, and the multitude of one-off case study projects. Parity Projects have continued the development of tools and methodologies of increasing sophistication – an SME doing the task that should really have been entrusted to a state-funded 'Manhattan Project,' addressing what the most recent parliamentary committee report has termed the need for retrofit to be put 'on a war footing.'

Starting in 2005, with experiments Parity's founder carried out in his own home, using a variety of insulation materials, Parity Projects has undertaken studies of increasing size. Throughout, they have emphasised that even similar house types can have very different energy consumptions, largely due to the variety of household sizes and lifestyles. This points to the difficulty of a 'one-size-fits-all' approach. For example, when consultants were called in to advise on the retrofit of part of Thamesmead, they found that multiple occupation was so extreme that only a mechanical ventilation system offered the chance of good indoor air quality. But the rate at which people move, and households change, highlights the problems posed by precise and inflexible solutions.

One of the first area-wide analyses carried out by Parity Projects was in 2009 for the Borough of Merton (Parity Projects 2009). The aim was to establish, within the housing stock in the borough, the potential for carbon savings, to identify the most favourable technical solutions in terms of suitability for each dwelling type, and to estimate the cost of installing packages of measures. Every dwelling in the borough was subject to a visual inspection that identified those characteristics with energy implications, their age, construction, number of storeys and bedrooms. In addition, a number of more detailed surveys were conducted into generic archetypes. By this time, Parity had developed its own domestic carbon modelling software based on the BRE's BREDEM, which enabled the outline information to be turned into detailed models incorporating, for example, wall and window areas. From this, a long list of potential measures was aggregated for all the borough's stock, with estimated payback times, total carbon savings, and the percentages of the stock for which each measure would be applicable. By considering the extent of applicability and cost effectiveness, packages were selected according to their payback periods. In addition to technical items, the study also estimated the effect of changes in individual behaviours that cost nothing but can be highly effective if behaviour change comes about. The conclusion was that these behavioural measures, plus solid wall insulation, were likely to be the most effective strategies if savings in the order of 60–70 per cent were to be achieved, but that savings larger than around 35 per cent would be difficult in the absence of household behaviour change (Parity Projects 2023).

In the intervening years, Parity Projects have enlarged on these methods of stock profiling and scenarios for achieving energy efficiency. But it has become apparent that action has to be based in communities, and that a movement for change is needed, not to mention the 450,000 tradespeople necessary to undertake the work. Most recently, Parity has analysed every address in Essex and arrived at a total retrofit cost for the county of between £13 billion and £17 billion according to the measures adopted (Parity Projects 2022).

So now Parity Project's emphasis is on implementation, an idea to draw together all the elements and trades required for successful retrofit into a 'one-stop-shop,' capable of replication between cities, called RetrofitWorks (RetrofitWorks 2023). All the diverse factors from the initial energy audit, financing, project management and monitoring of final performance come under the RetrofitWorks umbrella, to create a process that can inspire trust through achievement of the best results with the minimum spend. Retrofit continues to labour under all the difficulties it has faced since the beginning, from a lack of customers, who without knowledge of the alternatives and benefits find the available offers unappealing; to tradespeople with little confidence in the retrofit market; financial institutions with a similar lack of confidence; and local councils and charities who want to promote change but struggle with all of these impediments. To draw together all of these disparate groups, RetrofitWorks aim is for a portal, supported by subscription from a vetted supply chain, to provide online tendering, access to finance and grants, and to provide training and instruction. The hope is that by creating a route for customers into the process and smoothing the local supply chain from design and contract, to installation and feedback, retrofit can become a reliable proposition (around the country other attempts are being made, such as Carbon Co-op in Manchester, and Dm's in Edinburgh).

The first application, in 2019, was for Cosy Homes Oxfordshire (CosyHomes 2023). Its website invites homeowners to enter their postcode, thereby giving them access to a database, derived from surveys and Energy Performance Certificates that summarises the energy efficiency and performance of comparable properties in their locality. The model called 'plan builder' builds a draft retrofit plan which can be edited to suit individual needs; for example, whether the intention is to reduce bills or lessen carbon impact, it then suggests a menu of costed alternatives. Supplementary videos about topics such as insulation, and different specifications for windows, provide background information. Once the plan is submitted, the next steps get underway, first a complete building assessment, and then the involvement of collaborating Retrofit Coordinators and contractors. The latter are invited to become members of the RetrofitWorks cooperative with the attraction of TrustMark accreditation and access to ongoing training and development.

The project was funded by government grant for a period of 30 months from the end of 2018, after which it has become self-funding as a result of 10 per cent of the value of the works going back to the scheme.

RetrofitWorks is an innovative approach, to break the retrofit logjam through smart marketing, overcome consumer resistance to having to deal with builders whilst having little idea of what to want or expect, and ensuring the process is smooth by introducing contractors with appropriate training and skills. But will it overcome the longstanding problem of disruption to households for the distant benefit of payback over a number of years? Homeowners, living in some of the country's most poorly performing dwellings (let alone the private rental sector), have proved particularly hard to entice since the costs have to be met up front; rather than, as has been attempted in the social sector, by repayment through energy bills. So many housing innovations are subject to the vagaries of the economic cycle, RetrofitWorks remains an important work in progress.

Note

1 See for confirmation: Dow, B. 2020. "Profession's Failure to Embrace Retrofits Is a Macho Hangover of Modernism." *The Architects' Journal* 20th January.

References

Brand, S. 2010. *Whole Earth Discipline*. London: Atlantic Books.
CosyHomes. 2023. 'CosyHomes Oxfordshire.' Available online: https://cosyhomesoxfordshire.org/ (accessed 31 May 2023).
Fisher, M. 2012. *Capitalist Realism, Is There No Alternative*. London: Zero Books.
Jacobs, J. 1961. *The Death and Life of Great American Cities*. New York, NY: Random House.
Jameson, F. 1991. *PostModernism, or, The Cultural Logic of Late Capitalism*. New York, NY: Verso.
Lent, J. 2017. *The Patterning Instinct: A Cultural History of Humanity's Search for Meaning*. New York, NY: Prometheus.
Malm, A. 2018. *The Progress of This Storm: Nature and Society in a Warming World*. New York, NY: Verso.
Merz, J., P. Barnard, W. Rees, D. Smith, M. Maroni, C. Rhodes, J. Dederer, N. Bajaj, M. Joy, T. Wiedmann, and R. Sutherland. 2023. "World Scientists' Warning: The Behavioural Crisis Driving Ecological Overshoot." *Science Progress* 106 (3): 1–22.
Morton, T. 2018. *All Art Is Ecological*. London: Penguin.
Parity Projects. 2009. *Carbon Assessment of Domestic Housing in London Borough of Merton, version 1.0*.
Parity Projects. 2022. *Cost Effective Solutions*, ENE Workshop 12th December.
Parity Projects. 2023. 'Housing Data Analytics.' Available online: https://parityprojects.com/ (accessed 31 May 2023).
RetrofitWorks. 2023. 'RetrofitWorks: Building Efficiency Together.' Available online: https://retrofitworks.co.uk/ (accessed 31 May 2023).

4 Sustainable Retreat

'James Lovelock, the visionary engineer who first conceived of the earth as the self-sustaining organism he called Gaia, offers a sombre prophecy for humanity.' Humankind, he writes: "… would wake up one day to find that (we) had the permanent lifelong job of planetary maintenance engineer".

Sustainable retreat is a concept developed by James Lovelock: '… in order to define the necessary changes to human settlement and dwelling, at a global scale, with the purpose of adapting to global warming thereby reducing its expected negative consequences on humans.' Lovelock thinks: '… the time is past for sustainable development; we have come to a time when development is no longer sustainable. Therefore, we need to retreat' (Lovelock 2011).

The concept of sustainable retreat emphasises: '… a pattern of resource use that aims to meet human needs with lower levels and/or less environmentally harmful types of resources.' In terms of technological development, building construction is a laggard. Housing retrofit is even more disadvantaged, the Retrofit for the Future competition and its meagre follow-up the Retrofit for the Future (WHR) competition that sought to reduce costs, were 'toes in the water' for a much-needed programme of research and development; one aspect of the New Manhattan Project that should have been launched many years ago to find new ways to reduce energy and resource consumption. Retrofit measures ought to be like a suit of clothes that can be patched, mended and passed on as occupancy and climate change, both of which are unpredictable. At present, the technical means available aren't sufficient for the long-term task.

Mark Lynas … believes that the Anthropocene could turn out to be a wonderful era for humanity. "As scholars, scientists, campaigners, and citizens" runs his Ecomodernist Manifesto, "we write with the conviction that knowledge and technology, applied with wisdom, might allow for a good or even great, Anthropocene. A good Anthropocene demands that humans use their growing social, economic and technological powers to make life better for people, to stabilize the climate and protect the natural world". This, say the believers in a bad Anthropocene, is madness. They see ecomodernism as a humanist superstition. They claim that, like religions of the past, it is a way of pacifying the people, preventing them from acting to save the planet from the rampages of global capitalism.

(Lovelock 2020, 67–68)

DOI: 10.4324/9781003564997-6

Technology (alone) is not the answer

> There aren't any magic technologies so we have to face facts. As we stand, an 80 per cent cut in emissions means an 80 per cent cut in energy use and future supplies from renewable sources can only reduce this target by a small fraction. For the three main areas in which we use energy, this target means (1) travel 80 per cent fewer miles than we do today; (2) turn on the heating for 12 minutes of each hour of current use; (3) reduce our purchasing of new materials – such as offices, houses, roads, cars, furniture, clothing and paper – to a fifth.
>
> (Attwood and Cullen 2011)

It's apparent now that there is no silver bullet, whatever breakthrough technologies may arrive in the near future won't be able to be scaled up in time to meet the 2050 zero-carbon deadline, as most recently confirmed by the March 2022 IPCC Report. In recent times, the weight of cars has been increasing, as has the temperature to which houses in the United Kingdom are being heated. So the climate crisis isn't just a technological problem. Despite the technical efficiencies achieved to date, the behavioural and societal pressures remain unabated.

For architecture, though, it is at these moments, when the cultural change of gear is grinding, that a new architecture emerges. In the societal cataclysm that followed World War I, manifestos were in abundance; the propulsion towards a clean new world of speed and technological excitement animated Futurism, Constructivism and Towards a New Architecture. Le Corbusier's advocacy of a clean sweep through the inherited city found its way to the mainstream, and enaction following World War II, as the unequalled public housing programme.

The cartesian grid of Modernism is imbued with the Platonic division between mind and body, which became the Christian divide between the human soul and the biological human. The status afforded the logical brain, assumed superior to the state of nature, fostered the Enlightenment, and the West's propulsion to colonial dominance, aided by science, power and weaponry. The same urge to dominate can be felt in Corbusier's sweeping away of the 'fungus' of Parisian café street-life. It is the same division that relegated women and their assumed reliance on intuition, empathy and care to unimportance. It is time then to re-assess what Modernism has given us that will guide the way towards a new architectural chapter, and other parts of the inheritance – the machine aesthetic, for example – that will form part of history but not find inclusion in the way forward. Our attitude to the city, as with our attitude to nature, will need to change from one of male domination to one of female nurture.

The future slowdown

It could be thought that trends within some of the world's statistics will before long undermine the economic system that has absorbed architecture into itself. Not only the gradually reducing growth in GDP (in both the West and China) but also the slowing rate of population growth (including Africa) may issue in a non-growth society. Unfortunately, the one trend that continues to grow exponentially is the rate of carbon dioxide emissions.

> We are now so used to the inevitability of technological change that we find the simple fact of technological slowdown almost impossible to accept. But most new products released in the last decade have involved little more than tinkering round the edges. As societies across the world have become richer, each small incremental change to our quality of life has mattered less and less. There are now clearly diminishing returns to technological advancement, and soon this fact will be so commonplace that you will be bored when it is mentioned.
>
> What changed most abruptly, immediately before the slowdown, was a decline in extreme scarcity and acute hardship for our species… As our material life has become easier, as the slowdown progresses, and as we are less easily diverted by new technological toys, we have become increasingly cognisant of the harms we are causing. We see not only the collateral damage and the externalities, but also the inner harm to our psychological and emotional worlds. We have softened, and so have the class and social relations among us. Less is repressed. We are now less violent. We have become a little less hierarchical. We are much more aware of all that we harm or destroy.
>
> (Dorling 2021, 182)

The idea of progress

Danny Dorling suggests that capitalism may just end with a whimper. As productivity and growth in GDP slow to nothing, led by the Western world, so will the rate of technological and social change. Ahead of the pack is the Netherlands that is running out of space and workers, whilst enjoying the shortest of working weeks at 30.3 hours, and where 60 per cent of the workforce choose to work part-time. For example, in terms of space, Dutch farming is so efficient that it is the world's second-largest agricultural exporter, but its 15 million cows and pigs live next to protected environments resulting in nitrogen emissions that exceed EU law. The Netherlands, despite hosting Europe's largest port, and being one of the world's most globalised economies, may soon be finding the limits to growth.

How to deal with the remnants of turbo-capitalism, the accumulation of debt and the piles of architectural detritus? But whereas capitalism has catered for limited human needs, stability may foster wider satisfactions. Tim Jackson in *Post-growth* identifies 'flow' the point of perfection reached, for example, by championship tennis players whose game is so developed as to appear instinctual, a perfectly attuned string quartet, or architecture honed to conceptual realisation at the level of detail. It suggests a society with work/life balance where these skills and facilities will be given the latitude to flourish. A new renaissance perhaps where a plaster and vegetable pigment fresco will be the desirable home addition rather than the latest Ferrari?

> ....the theory of deep ecology and with the Gaia hypothesis – embodies a challenge to the ruling worldview of the age, which is a sort of scientific fundamentalism allied with liberal humanism....The idea of progress is rightly anathema to the most reflective Green thinkers, one of whom has stigmatized it as expressive of 'the anti-way', the way downwards, to entropic disorder and final extinction... There can be no doubt that the project of a social order that does not rest on the prospect of indefinite betterment creates problems for policy that have as yet been barely addressed by conventional thought... A universal stationary state may well be a utopia, but it is a better measuring-rod for attainable improvement in the human lot than the wholly unrealizable fantasy of infinite growth.
>
> (Gray 1993, 314–332)

Modernism and the natural world

> In his popular best-seller, Last Child in the Woods: Saving our Children from Nature-Deficit Disorder (2005), Richard Louv diagnosed Western consumerism provocatively as a culture suffering from a nature-deficit disorder, suggest that a lack of care for the environment is the direct result of a lack of attention given to the environment... Though directing attention to the environment may be hard work for one who has grown accustomed to distraction, paying attention as such requires no special tools. As Timothy Morton writes in *Being Ecological,* the problem with climate change is not that it is difficult to solve. If anything, because we are the tools, the solution is 'too easy'... Considered as the principle of agency in the Anthropocene, what precisely is attention? We might define attention as a way of looking at the world with concentration and deliberate effort, but also by care ... The reason why Louv, for instance, wants his readers to pay more attention to the environment is because he hopes they will then be more likely to care about and for the environment. Once again it is a case of attention generating a positive feedback loop. The habit of attending to the environment becomes second nature, or rather, transforms and extends nature, since the more attention one gives to the environment the better adapted one then becomes to it.
>
> (Kotva 2018)

It has become usual in current discussion about the Anthropocene to investigate our current situation in terms of the human inheritance, particularly the Western philosophical tradition. Casting back to the agricultural revolution and the first cities, the social and class organisation that structured the new human settlements, superseded the hunter-gatherer way of life. These arguments are frequently coined in terms of loss, the exchange that was made for the totality of experience that was a fulfilled existence in return for stability, the repressive aspects of power and the paucity of experience that followed from Western dualism. That divide between body and mind, and the primacy given to the latter, derided all that was biological and, by extension, the natural world. Similarly, attention, the notion that thought and matter do not have an independent existence, but that attention can exert change, in the way, for example, that the brain adapts to learning to ride a bicycle, or learn a foreign language. Herbert Marcuse *in Eros and Civilisation* (1972) traces the denial of sensual pleasure in the establishment of a class-structured society. Tim Jackson in *Post-growth* (2021) suggests that capitalist consumerism exploits the human instinct for novelty whilst denying other psychological needs, in the same way that Gross Domestic Product is a poor measure of the totality of human existence.

> 'We are not like the butterfly, whose generations are unknown to each other; we are a familial and historical species, for whom the past must have authority (that of memory) if we are to have identity, and whose lives are in part self-created narratives, woven from the received text of the common life.'
>
> (Gray 1993, 309)

Architecture as a belief system

> It may indeed be that the Gaian vision, being free from the anthropocentrism which privileges humans in the universe and which even models the universe on humans, is the most appropriate antidote to this malady of the spirit that parades as enlightenment.
>
> (Gray 1993, 383)

Architects are finding themselves on the wrong side of history. Nature, the environment and the sanctity of the world are becoming the elements of a new secular religion. Putting aside the elevated status accorded to man by the old religions, religions too are attempting to reposition themselves in relation to the new eco-reality. After World War I, architecture became modern, it moved away from nineteenth century battles about a preferred golden vision of the past, Classical or Gothic, the ancient past or the medieval, and set its face to the future. Architecture became aligned with progress, positivism and science, and in turn with the ever-upward march of economies and markets.

> Architecture and urbanism are always concerned with the future. Restoring the future to good condition does not mean more of the same, largely developer-driven and capital-centric architecture and urbanism, but rather a long-term commitment to planetary care based on human and non-human liveability. We see an ethics of care as the most important perspective for an architecture and urbanism in our times of catastrophic ruination.
>
> (Fitz and Krasny 2019, 12)

Case Study: Link Road, Birmingham

Across four days in July 2022, the Retrofit Reimagined neighbourhood festival and conference was convened in Birmingham, involving a number of mostly young, activist organisations, many of the participants described themselves as 'ex-architects'; they had found a new way of making a difference. Rather than the usual fate of young architects, as conscience-stricken CAD monkeys, in this context, they found a role as valued members of the community team and part of the conversation within an array of new voices. Rather than the usual dry conference format, here the discussions were held in an open tent, welcome to all.

Amongst those presenting: Architects Climate Action Network (ACAN), Insulate Britain, the Anthropocene Architecture School and Retrofit Action for Tomorrow. The local community platform Civic Square organised the event, cognisant of all the difficulties resulting from poor housing, not just energy issues but health problems, welfare, inequality, racial and climate justice. But also, they could perceive, from the recent development of one-stop-shops and neighbourhood campaigns that the ever-faltering rate of government-sponsored retrofit might be energised from within local areas, street-by-street retrofit being the goal.

The tone of the event was set by the introductory speaker from ACAN (Civic Square 2022a), who described herself as 'not an architect any more.' Through awareness campaigns, ACAN is conversing with a range of audiences in the cause of radical change: 'If people were allowed "free-thinking" they would have more power than they realize.' The emphasis of the event was on local action to make retrofit happen, ownership of the process by communities. The speakers were an inclusive group – practitioners, community advisers, teachers and health specialists.

Dark Matters Labs (Dm) presentation (Civic Square 2022b) compared the possible cost of UK retrofit of around £1.45 trillion with the national debt of £1.9 trillion; civic activism is necessary to face a scale of challenge larger than the post-war housing programme, the centrepiece of which is the home. Climate change is, he noted, a symptom of our failure to relate to the world. His presentation looked at the big picture: To undertake current forms of retrofit, the country hasn't a hydrocarbon allowance sufficient to mine the materials to do the work, and there is not the available labour force. How can we make decisions as a society when democracy needs reform? A foundational shift is necessary that this generation will need to bring into effect; the general population understands the problems intuitively 'at the scale of what it means to you today.' We are nowhere close to solving any of the big issues; global warming, for example, is happening a lot faster than predicted. How to transform a whole nation's housing, not just from a carbon point of view, but with the opportunity to make better places, thereby rebuilding our basic living environments, both streets and neighbourhoods? Conversations need to be held in a different way to start rebuilding everyday politics. Street-by-street retrofit will be the equivalent of post-war social housing in terms of reimagining the idea of housing, rebuilding it in a new way, by communities themselves through a public civic partnership, government in co-production with communities; an opportunity to rebuild our relationship with the world.

Dm are deeply engaged with street-by-street retrofit, the street being perceived as a 'unit of organisation' that, through collective financing, can initiate not only building retrofit but improvement in the public realm, and community facilities – a total retrofit. Rather than merely emphasising decarbonisation, there needs to be recognition of the wider benefits of retrofit, in terms of improved local skills and services, which haven't been factored into the very narrow definition of performance that has dominated the debate so far. The transformation is foreseen as happening within communities by building trust and enlisting local capabilities, local supply chains and local networks of knowledge, and promoting adaptive rather than one-size-fits-all solutions.

Dm continued (Civic Square 2022c) by listing the problems of retrofit: Over consumption of energy, the housing crisis, health and social isolation, the necessary transition to a different sort of economy, the crisis of trust in government and democracy. These aren't merely technical problems to do with heat pumps and solar panels, but have wider repercussions for the life of the street. They have been researching the problem of how to finance the retrofit of Link Road and how their findings could impact a national plan. The post-war housing model won't work because retrofit is too complicated. Green Homes Grants leaflets, for example, are delivered door to door and go straight from the doormat to the bin. These initiatives, designed purely around funding, encourage cut-price construction with concomitant knock-on effects, the emboldening of cowboy builders, for example. Link Road has been adopted at a test bed. Local residents have been encouraged to get involved, to join together and explain why they care about their street, through events, a walking tour with a thermographic camera, and presentations to schools. Retrofit assessments have been carried out, the results being fed back through collective sessions on the street. The next step is to research funding, to find what money is, or could be, available. Then, to form a street association that will become an energy collective, using PV installation as a way to build revenue. The plan is for it to develop into a retrofit service company, a non-profit club exchanging solar electricity for income; local contracting firms to join as members. The capital cost for Link Road's retrofit they estimate at between £1.1 and £1.2 million. Even including available grants, through the local area delivery scheme and heat pumps scheme, and solar roofs revenue, there will nonetheless be a shortfall of about a third, so the finances at present don't add up. Asking homeowners to fund the rest is an overly large additional cost to impose on household expenses.

Finally, Dm enlarged on the problem of knock-on effects, the cascading costs of bad health, and the risks posed by dangerous environments. The costs of bad housing are met by a variety of agencies, public bodies and government. Using Link Road as a testbed, the aim is to identify what makes a street a good, or not good place, to determine the feasibility of street-by-street retrofit, and to identify how many other streets and communities the principle could be applied to. They are seeking to quantify the social and health benefits so communities can see the costs and trade-offs as an aid to making decisions along the way to a better future.

Civic Square (Civic Square 2022d) wound up the event with the maxim: 'from technology to social movement,' Civic Square's role being to reframe the conversation. Democratic systems are not up to the purpose, everything is subsumed to the political economy of neoliberalism. How did we get stuck? At the most recent, ineffectual, COP there was enormous pushback on the street by indigenous peoples and direct-action groups. What limits the ability of the young generation to organise? Is it precarity and time poverty, those time and energy deficits that deter joining in protest? Retrofit becomes a question of how to get everyone involved, rediscovering concern for the 'commons' and its governance. Architects see retrofit as a technical challenge but it's a technical task that depends on social organising for implementation. Civic Square is using Link Road as a vehicle to start conversations about local governance. The question being, can housing retrofit be the mechanism to build a social movement and launch 'hyperlocal governance'?

References

Attwood, J., and J. Cullen. 2011. *Sustainable Materials – With Both Eyes Open: Future Buildings, Vehicles, Products and Equipment – Made Efficiently and Made with Less New Material (without the Hot Air)*. Cambridge, UK: UIT Cambridge Ltd.

Civic Square. 2022a. 'Retrofit Reimagined: Why We're Here', YouTube, 14 July. Available online: https://www.youtube.com/playlist?list=PLgIJxOKjWOpigXk97jKPuBLf5fsfGNWjb (accessed 15 September 2022).

Civic Square. 2022b. 'Retrofit Reimagined: Indy Johar (Dark Matter Labs)', YouTube, 14 July. Available online: https://www.youtube.com/playlist?list=PLgIJxOKjWOpigXk97jKPuBLf5fsfGNWjb (accessed 15 September 2022).

Civic Square. 2022c. 'Retrofit Reimagined: Jack Minchella + Calvin Po (Dark Matter Labs)', YouTube, 15 July. Available online: https://www.youtube.com/playlist?list=PLgIJxOKjWOpigXk97jKPuBLf5fsfGNWjb (accessed 15 September 2022).

Civic Square. 2022d. 'Retrofit Reimagined: Closing Panel with Charlie Edmonds, Gavin Rogers, Phil Beardmore + Scott Hewer', YouTube, 16 July. Available online: https://www.youtube.com/playlist?list=PLgIJxOKjWOpigXk97jKPuBLf5fsfGNWjb (accessed 15 September 2022).

Dorling, D. 2021. *Slowdown: The End of the Great Acceleration and Why It's Good for the Planet, the Economy, and Our Lives*. New Haven, CT: Yale University Press.

Fitz, A., and E. Krasny. 2019. *Critical Care: Architecture and Urbanism for a Broken Planet*. Cambridge, MA: MIT Press.
Gray, J. 1993. "An Agenda for Green Conservatism." In *Gray's Anatomy*. London: Routledge.
Jackson, T. 2021. *Post Growth: Life after Capitalism*. Cambridge, UK: Polity Press.
Kotva, S. 2018. *Attention in the Anthropocene in Political Geology*, Active Stratigraphies and the Making of Life. Available online: https://www.geog.cam.ac.uk/events/politicalgeology/
Lovelock, J. 2011. '*Looking into the Future' Quoted in the RIBA Journal*,' April: 35.
Lovelock, J. 2020. *Novacene: The Coming Age of Hyperintelligence*. London: Penguin.
Marcuse, H. 1972. *Eros and Civilisation: A Philosophical Inquiry into Freud*. London: Abacus.

5   Caring Architecture

> Architecture in its broadest sense provides shelter indispensable to the continuation of human life and survival. This is evidently a form of care. Yet historically, architecture has not been considered a form of caring labour. Despite this fundamental function of architecture to provide protection for humans from sun, wind, snow or rain, and to give the support necessary for maintaining the vital functions of everyday living, the idea of the architect is linked to autonomy and independent genius rather than connectedness, the architect being an artist, traditionally gendered male, has been most influential over centuries, the notion of the architect being a carer, traditionally gendered female and considered menial labour performed by racialized other, has been completely absent from the discourse on architecture…we suggest that caring be viewed as a species activity that includes everything that we do to maintain, continue, and repair our 'world' so that we can live as well as possible. That world includes our bodies, ourselves, and our environment, all of which we seek to interweave in a complex, life-sustaining web.
>
> (Fitz and Krasny 2019, 33)

In the new context, this is very different from the need to clear and rebuild that was the imperative after 1945. Now we are left with the remnants of excess but within a climate in crisis the new imperative is to make do and mend, to darn and mend the fabric of the city. The city now becomes understood as a complex ecosystem in its own right that, like a natural ecosystem, can only be altered with full understanding of the many interacting vectors – social, economic, material, organisational and ecological that need to be anticipated and addressed. This is not simply a need to build without plastic, to conserve energy, to achieve a circular economy, but the theoretical underpinning of a new way of making architecture.

Scale up this model to include all the flora and fauna of Earth and you have the system I have called Gaia. In fact. you cannot actually scale it up because the system is too complex, so complex in fact that we are nowhere near fully understanding it. Perhaps it is hard to understand because we are an intrinsic part of it. But also, I suspect, it is because we have been too reliant on language and logical thinking and we have not paid enough attention to the intuitive thinking that plays such a large part in our understanding of the world… Human civilisation took a bad turn when it began to denigrate intuition. Without it, we die. As Einstein said "The intuitive mind is a sacred gift and the rational mind is a faithful servant, we have created a society that honours the servant and has forgotten the gift". Perhaps it happened because women's insights were rejected.

(Lovelock 2020, 13–20)

If we are heading towards a post-growth world which heralds an era which is also post-progress the current trend towards slowing rates of change will become the norm. What is suggested by Tim Jackson et al. is that this could be a gain for culture. Rather than the limited expectations of capitalism, endless competition based on relentless testing and assessment, the need for growth in GDP, material possessions and the latest novelty, different criteria could prevail. A change in timescale and decision-making for the long term would require architecture to envisage buildings as permanent components of the urban ecosystem. Like shrines in Japan that remain unchanged throughout the centuries but are cared for and retrofitted piece by piece as maintenance cycles demand, or medieval cathedrals in Europe which are subject to the same level of care, buildings generally will enjoy a similar regime.

Caring architecture will not be the same thing as sustainability, as important as that movement is. Sustainability began in the 1980s as an attempt to make architecture more sensitive to its environmental impact. But as it was institutionalized, its standards became about things: it focused more on the materials used and it has been more successful in measuring what goes into a building than in monitoring the ongoing effects of sustainable building. Because care emphasizes processes and relationships that extend back and forward through time, and concerning all of the created relationships, applying care theory to architecture would involve making a fundamental shift in perspective: care does not view the completed 'thing' -building, park, city, zone etc – as its object. It starts instead from responsibilities to care, not only for this 'thing', or its creator, builder, or patron, but for all who are engaged in contact through this thing.

(Fitz and Krasny 2019, 28)

The future role of the architect will be to build sparingly

The conclusion of Andreas Malm's *The Progress of This Storm, Nature and Society* (2018) after tracing the emerging outline of life on a warming planet, suggests our principal task over the next centuries should be stabilising the climate, to achieve the restoration of nature as a free-functioning autonomous entity, able to function without threat to humankind. For architects, the path is clearly to make no more incursions into the natural world, which needs to be left to once again find its own equilibrium. The raw materials of steel and concrete, like fossil fuels, need to be left in the ground. The city *as found* becomes the architect's sphere of activity, one of nurture, only building that which is absolutely necessary and only with materials re-used and locally sourced. The extent of building is then to fill vacant sites and increase density, thereby growing back the stock of publicly owned housing;[1] or to replace buildings that are so deteriorated as to be beyond improvement; or to reanimate deprived neighbourhoods; or to utilise available structural capability to add extra storeys (for which the Lambeth Skyroom project (Woodger 2021) is a prototype). With GIS, it should be possible to build a national atlas by use type, construction and energy use and their potential for renewables. This has already come about in Utrecht where a 3D virtual model has been built, which classifies buildings by their energy-saving potential related to retrofit costs. The amount and location of new construction becomes a matter of debate, but also part of a national retrofit programme needs to be in harnessing data to provide the basis for evaluation, a new Domesday Book, one of the many areas of retrofit knowledge that are sorely lacking and badly needed.

The closest we have come is The National Energy Efficiency Data-Framework (NEED) that was set up to give a clearer picture of energy use and energy efficiency. The data relates to gas and electricity use with information on energy reduction measures installed in homes as a result of the various grants that have been available over the years. It includes data about property types and household characteristics but not at the granular level, which could be achieved using GIS. A great step forward was made in 2020 with the launch of the London Building Stock Model that contains three-dimensional information about every building in London, including the uses they are put to, and their Energy Performance Certificates, a tool that needs application across the whole of the United Kingdom. Though a true Domesday Book would also incorporate a 'digital twin' (Susy n.d.) 'whole house' retrofit plan for each address including the building's suitability for renewables (according to their modelled energy output compared with household demand) and the carbon implications of work-staged maintenance upgrades being anticipated by lifecycle analysis. This approach would parallel the 'golden thread,' advocated since the Building Safety Act, whereby every new building will have a database recording its history from construction onwards.

Building little implies making good what we already have i.e. retrofit

Within this inclusive perspective, retrofit is central to the purpose of an architecture that is more than the quest for visual perfection. Studies that unfortunately have only been carried out at a small scale, illustrate that a national retrofit programme will only achieve the necessary reductions in energy consumption if occupants are engaged with the programme and their retrofitted homes. The human and temporal dimensions challenge Modernisms limited agenda. Although it's only 10,000 years since the agricultural revolution, the beginnings of social order and religion, our time scales have become ever shorter. 2050 is now the hoped-for zero-carbon world, though that won't mark an end to species extinction, the once multitudinous insects upon which our existence depends. Instead, in a post-growth world, time scales will need to be radically extended. For architects, how to make long-life/loose fit happen so retrofitted buildings can adapt flexibly to unforeseen human and climate conditions will be the great challenge.

> 'We have to be careful what we humans design, because we are literally designing the future, and that future isn't in our idea of the thing, how we think how it will be used and so on – that's just our access mode. The future emerges directly from the objects we design.'
>
> (Morton 2018, 209–210)

Whilst the debate rages on, fortunately for architects there is emerging clarity. To discard Modernism entirely would be to not only turn our backs on the problems Modernism has created but also to dismiss our means for addressing them. A reassessment of Modernism differentiates those aspects that belong to the past and those that find a way into the future. A slowing world without progress will also be one to care for ecologies both natural and organic or inanimate, including the urban ecology. Throughout recent times, it has been usual for architects and particularly environmental engineers to construct their own assessment tools based on in-house data that is commercially sensitive and therefore unavailable for public use. Decision-making that needs to be made at increasingly large scales, the scale at which weather systems operate, needs all this information to be made available for ecological engagement. This is the scale of the Internet and social media; the world is getting smaller, which makes a post-growth world, and its architecture, a realisable utopia.

Note

1 So key workers can live affordably close to where they work. A case in point are the empty rooms above shops, which attracted the attention of Gordon Brown, but which remain prevalent. Now many of the shops themselves are vacant the potential for retrofit is that much greater.

References

Fitz, A., and E. Krasny. 2019. *Critical Care: Architecture and Urbanism for a Broken Planet*. Cambridge, MA: MIT Press.
Lovelock, J. 2020. *Novacene: The Coming Age of Hyperintelligence*. London: Penguin.
Malm, A. 2018. *The Progress of This Storm: Nature and Society in a Warming World*. New York, NY: Verso.
Morton, T. 2018. *Being Ecological*. London: Pelican.
Susy. n.d. Available online: https://susy.house/ (accessed 2 April 2024).
Woodger, M. 2021. 'Affordable Room up Top', *RIBA Journal*, April: 32.

Part II
A Brief History of Retrofit

6 Size of the Problem

The problem is the unruly confusion of the built environment. Sixty per cent of the UK's energy is consumed by its 28 million homes, some of the oldest building stock in Europe, probably 25 million need retrofitting. To achieve net zero, since 2021, the UK's Carbon Budget has included aviation and shipping emissions, now estimated in total to achieve more than three-quarters of the overall reductions necessary by 2050. The improvements required of the building stock will have to be more substantial yet, 1.7 million heating systems that run on fossil fuel are still being installed each year. The numbers are daunting, 430,000 retrofits will need completion by 2030. Because the rate of demolition and renewal is so slow, particularly in recent years, barely 20 per cent will have been replaced by 2050 and the rest will need to have been retrofitted at a cost of £12 billion per year! The building stock is of course far from homogeneous. Different house types consume different overall amounts of energy, and the split between the proportions devoted to space heating, water, appliances etc. differs with the numbers of occupants and their patterns of behaviour. For example, solar thermal panels may be a benefit to families with children who use a lot of hot water, but less so when the children grow up and leave home. Every dwelling becomes an individual design problem, but the flexibility of the typical English house that has ensured for many homes a useful life into old age, must not be compromised.

'Old age' is an appropriate epithet. Pre-1919 houses are around a quarter of the total but approaching half of the private rental stock, whereas nearly all local authority housing and the vast majority of other social housing are from after World War I. The *2020-21 English Housing Survey* (EHS 2021) records the improvements that have been made by capturing the 'low-hanging fruit' – windows being improved with double-glazing and the replacement of solid fuel heating with condensing boilers. In 2020, 87 per cent of English homes had double glazing, up from 74 per cent in 2010. Readily achieved increases in insulation, filling cavities in walls and loft insulation have also been made such that 49 per cent of cavity wall dwellings now have insulation fill, an increase of 2 per cent from the previous year. Overall, 52 per cent of homes now have cavity or insulated solid walls, compared with 41 per cent in 2010 and 39 per cent have at least 20 centimetres of loft insulation, up from 27 per cent in 2010. Consequently, SAP ratings have increased so

DOI:10.4324/9781003564997-9

only 10 per cent remain in bands E, F and G; now 87 per cent fall within bands C and D, compared with 61 per cent in 2010. The average SAP rating has reached 66 points, an increase of 21 points since 1996. But whilst the United Kingdom intends all properties to have achieved EPC grade C by 2030, with only six years to go, only 29% are achieving the goal.

The equivalent study carried out in 2006 (EHCS 2006) illustrated the extent to which improvements are made in line with current expectations for comfort and amenity; nearly 20 per cent of homes nationally had been extended, though less than 10 per cent had undergone complete refurbishment. Back then, a startling number did not conform to the standard of a 'decent home,' 35 per cent of the overall total and nearly half of the privately rented sector. In the meantime, matters have improved somewhat. In 2020, 13 per cent of social housing failed to meet the Decent Homes Standard, whereas 16 per cent of owner-occupied homes and 21 per cent of market rentals, that is, four million homes, were deemed substandard. The biggest problems being lack of comfort and excessively cold temperatures, though social housing has undergone considerable improvement in recent years, and consistent through the years of the survey, social housing is substantially more energy efficient than homes that are privately owned.

Modelled results in the *English Housing Survey 2020–21* showed that problems relating to damp have improved over the years. They can be caused by a variety of factors, including disrepair, over-crowding, insufficiently heated rooms and/or ineffective ventilation and poor thermal insulation. Reflecting these diverse causes, both old (overwhelmingly private) homes and newer social housing are amongst those most likely to have any damp problems. Damp is most prevalent in privately rented homes, but serious condensation and mould growth are equally common within social housing, as has been highlighted by press reports, yet again, in 2023. Whilst the BRE's 2023 Cost of Poor Housing Report (BRE 2023) estimates that improving the 65,000 homes with the worst damp and mould problems (category 1) would achieve £4.5 billion of societal benefit for an expenditure of just £250 million.

The statistics have much wider implications for the quality of life within communities; although the link between poor housing and poverty isn't straightforward, older and comfortless housing is found in both deprived and affluent areas. But there is a strong correlation between poor living conditions, health issues and problems within communities. The homes of those in poverty, social tenants, ethnic minority households and where there are children within poorer families are twice more likely than average to have insufficient upkeep and management of public and private space, and buildings, in their immediate neighbourhoods.

Post-millennium years of increasing global warming have highlighted the problem of summer overheating. In 2020, 8 per cent of residents considered that their homes, at least in part, got uncomfortably hot. This was most true in the case of owner occupiers, 9 per cent were concerned that an area of their home overheated, compared with 6 per cent of social housing tenants and 7 per cent of private rentals. Householders in dwellings built more recently reported an increased incidence of overheating. In 2020, 12 per cent of those living in homes built after 2003 were of the opinion that some part of their home was inclined to become uncomfortably hot.

The energy case for retrofit

In the United Kingdom, the motivations for the radical energy demand reduction that can be achieved by retrofit are many. Most clearly, it will be for the benefit of deprived communities in tackling fuel poverty and health inequality. As well as the extreme urgency presented by climate change, security of supply has become an issue since oil and gas come from increasingly unstable areas of the world. The roll-out of street-by-street retrofit, at a scale that will be necessary to meet these challenges, has never really begun. The logistical problems involved with decanting occupants, and retrofit's general disruption, are exacerbated by the huge variety of dwelling types. What is required is a national discussion, about, for example, the relative importance of heritage compared with the climate emergency. There are around half a million listed buildings in England alone, and there are hundreds of thousands within conservation areas. The policies for their retention are imposed with loosely defined rules that are locally applied and subject to a range of interpretations, and the wide variety of traditional construction types are poorly understood in terms of their capability for energy improvement. The Sustainable Traditional Buildings Alliance has listed the shortfalls in knowledge about energy performance: Traditional materials characteristics, occupant behaviour in older buildings, overheating risk, air quality issues and ventilation rates. All of these defy the usual analysis tools and are poorly served by the regulations.

Grosvenor Estates estimated in 2019 (Grosvenor Estates 2019) that 5 per cent of the nation's emissions due to buildings could be eliminated by retrofitting those that are listed and in conservation areas. On the one hand, many low-income households in older unimproved homes have high heating bills as a result, whilst some of the most expensive property in London is some of the worst-performing from an energy point of view. In Westminster, which consists mostly of ancient and protected buildings, it is estimated that up to 85 per cent of emissions are due to the built environment. Eighty per cent of Camden Council's social housing stock is within conservation areas, a lot of it very difficult to retrofit economically, mostly solid wall properties that are hard to improve other than with internal insulation (a problematic alternative). Of the quarter of homes in England built before 1919, many are in conservation areas. The competing demands of the carbon lobby and the heritage lobby ought to be addressed before fudged with potentially inadequate solutions being rolled-out wholesale. These are all big and important issues that require design ideas to prioritise the competing demands, and of course, the Building Research Establishment should long since have investigated new technical solutions and provided guidance. Instead of the scale of the 'Manhattan Project' that the challenges demand, just a few university departments, such as Glasgow Caledonian's, are trying to fill the void.

> Where there is coverage, new build tends to dominate, instilling the feeling that working on retrofit is second rate – that it doesn't require the same intellect or passion or creativity. But it does. Retrofit can be art, if it couldn't, I wouldn't be so interested. Of course, there will be retrofit projects where things are so constrained that all you can do is replace the boilers and plumbing… But then you probably won't need an architect. To my mind, architecture has become too narrowly defined. It's about space and image, but also about resources and the environment.
>
> (Prasad 2012)

The word 'retrofit' has already taken on such utilitarian and joyless associations, assumed a matter of boilers and plumbing, that the subject also needs an injection of architectural enthusiasm, as explained by Craig White:

> The retrofit market is enormous. However, no-one during my education as an architect said that the last half of my career would be taken up doing refurbishment! And certainly, I had no training of any sort. It's also difficult to sell the idea of a career in refurbishment to 18-year-olds deciding whether to invest £9,000 a year to become an architect in the longest vocational training route available. In the past, we did this by selling the promise of being able to design the future, not retrofit the past.
>
> (White 2011)

Problem #1 Determining the Outcomes

At present, the drive towards mass retrofit has a series of guides. On the one hand, the excellent PAS 2035 (BSI 2019a) defines a process to guide Retrofit Coordinators step-by-step, whilst the 2021 revision to Part L of the Building Regulations, marks a further ratcheting up of standards for fabric and equipment towards zero carbon. Whereas the requirement for works to existing building used to be on the lines of 'don't make matters any worse,' now retrofit has its own benchmarks. In addition, the beginnings of 'step-by-step' retrofit are defined by the increasing scope of collateral work that needs to be triggered when, for example, a window is replaced or a wall is improved. Interestingly, the regulations have now combined new build and retrofit, with the latter being a subset of the former, perhaps a sign that retrofit after years of inaction may finally have political traction. But the regulations are of course adapted from one revision to the next and accepting of compromise given the enormous variety of circumstances to be found amongst the inherited building stock. But this complication does make for imprecision, doubts about the degree of compliance actually being achieved are likely to be heightened when measures are deemed unnecessary if, for example, they have a simple payback period of more than 15 years, a metric that can only too easily be gamed.

The 80 per cent reduction sought by the government-sponsored Retrofit for the Future Competition (TSB 2009a; TSB 2009b; TSB 2014) was into the realms of the Passivhaus EnerPHit standard (Cotterell and Dadeby 2012; Taylor 2012; Traynor 2020). Passivhaus is a German building standard for new homes; it aims to be so energy efficient that a heating system is not required. The heat of the people, services and solar gain are sufficient to keep the houses warm; they are built very heavily insulated and very airtight. An insulated blanket is put around the building, walls, ground floor and roof, and an airtightness membrane as well, up in the loft is a mechanical ventilation system. In a Passivhaus, windows are kept closed throughout the winter, and air is supplied through ducts driven by fans. EnerPHit is a slightly reduced target for retrofit of existing houses, in recognition of the difficulty of achieving the same performance from existing buildings. There are 37,000 Passivhaus's in Europe and it has a growing following in this country. It is actively promoted by the Building Research Establishment.

To test the performance of alternative designs, the software for Passivhaus is called PHPP. Different forms of construction are possible to meet the Passivhaus and EnerPHit standards, but it is driven principally by the energy efficiency agenda and achieving comfort indoors. Critics point out though that making a very heavily insulated and airtight building a healthy place to live requires a ventilation system that works. Within Passivhaus there is only one ventilation system that is possible, which is called MVHR (Mechanical Ventilation Heat Reclaim). Wet, warm air is extracted from kitchens and bathrooms; a heat exchanger in the loft uses this heat to pre-warm supply air incoming from outdoors, which is then delivered to the living and bedrooms. There have however been a lot of problems with these systems, in terms of their installation, the noise of the fans, and because the filters have to be changed every few months; otherwise, the equipment doesn't work. But the big problem in retrofit is that it is very difficult, if not impossible, to simply install the ductwork when retrofitting two-storey houses. Whereas some existing houses, such as Victorian ones, are difficult to make airtight, newer ones can be relatively easy. The health implications of poor ventilation are considerable; in the United States, poor indoor air quality is ranked high amongst health threats; the United Kingdom has the greatest incidence of childhood asthma (associated with condensation and mould growth) of any country in Europe.

The majority of the Retrofit for the Future properties incorporated MVHR, but considerable problems were encountered. Ventilation has gone from being an overlooked requirement, not even able to be funded within the Green Deal (the since abandoned retrofit funding mechanism) to a major concern within PAS 2035.

Ductwork associated with heat recovery and ventilation systems has proved another major test. This is where a good design can fall down, as Beili from Paul Davis Architects attests: 'It was a massive job co-ordinating the ductwork, especially as the wet and dry rooms changed at the last minute.' And it's sensitive: these objects take up a lot of space. Bere Architects' end users temporarily halted the job when they saw the ducts going in, fearful of losing storage – a situation that took some careful negotiation.

(Young 2011)

Some architects remain sceptical of Passivhaus, Bill Dunster of ZED-Factory and Craig White of White Design are unsure if Passivhaus will work in the UK's mild climate, which makes mitigating electricity loads of mechanical ventilation harder to justify.

(There is a Passivhaus Rush 2008; Partington 2013)

The overriding question is whether the German Passivhaus standard is directly applicable to the United Kingdom. It's ironic that widespread endorsement of Passivhaus coincides with the diminished extent of building research now undertaken in this country. The increase in use of MVHR in houses is taking place when considerable efforts are being made to reduce the extent of mechanical plant used for other building types. For example, despite containing lots of equipment emitting heat, a natural ventilation system was installed in BSkyB's television studios. Natural ventilation is tricky; it entails moving air around at slow speeds, so large volumes are required, thus the size of BSkyB's ventilation chimneys. Within television studios, the heat produced by lights and cameras is the main issue. In housing, it's whether or not the fan power required for MVHR is justified in our relatively mild climate. Most certainly, doubts can be cast on the advisability of accepting a standard based on central European conditions. Germany, at the centre of Europe's continental land mass, has winter temperatures down to −20°C. The relatively warm sea surrounding this island maintains our equable winter average temperatures, which rarely fall below −5°C and average 10°C. But it is also a healthy buildings issue; the principal UK indoor pollutant is moisture, which leads to condensation, mould growth, dust mites and allergenic illness.

The National House Building Council (NHBC 2013a) is doubtful if Passivhaus principles are applicable in the United Kingdom for social, political and financial reasons. Social, because as opposed to Germany where many houses are self-built using high specification, well-detailed kits, in this country we rely on spec builders. In Germany, green politics have a higher profile than here; cities such as Frankfurt have Passivhaus-compliant construction written into law. And financial, since the additional cost there is only 3–8 per cent, which can be met with a government loan; whereas only one-offs have been constructed here so far, resulting in a lack of reliable cost data, although that situation is changing with time; EnerPHit projects are still, however, at an early stage in the United Kingdom.

The defects within retrofit schemes over recent years have resulted in ventilation becoming top of the agenda for PAS 2035. We've a history, in this country, of embarking on programmes driven by overriding needs that then fall foul of unforeseen side-effects. The high-rise housing of the 1960s, now largely demolished, was driven by the need to provide housing at speed, but the resulting damp and mildew-ridden blocks were unhealthy places to live. The Green Deal and the Retrofit for the Future Competition were concerned only with energy and carbon targets but risked future retrofitted homes being subject to similarly unforeseen side effects.

Gradually, if too gradually, the requirements of Part L are heading their way towards being very low energy. If building regulations are of necessity hedged around with provisos, the alternative approach, of absolute certainty, is taken by the EnerPHit standard. Slightly reduced from Passivhaus for new build but using the same approach – a lot of insulation plus an airtightness limiting value of 1 ach (at 50Pa) and a target of .6 ach. There is recognition though that whole-house retrofit is often impossible, and that there are limiting factors in existing buildings that can make airtightness difficult to achieve and cold bridges hard to eliminate. Consequently, a step-by-step approach has been introduced so a first certification can be made after only initial steps have been completed. Passivhaus products and tools are available to support the process, which is envisaged as either the sequential retrofitting of elements, walls followed by windows plus airtightness, ventilation etc., or the complete retrofit of whole sections of the dwelling. Perhaps the most valuable aspect is that each project proceeds from an overall plan, which can be entered and assessed within PHPP. This database could conceivably grow until the wide variations within the UK building stock could begin to be understood. The weakness is of course that first good intentions aren't conclusive. EnerPHit certification intends an outcome where actual performance, and predicted, coincide thus avoiding the 'performance gap,' each project having the quality assurance of using only approved materials and products. Inevitably though, a method that allows work to proceed at a pace tied to the availability of finance and willpower may, in the real world of compromise, never achieve the initial end-in-view.

Nonetheless, this is exactly what mass roll-out of retrofit needs, each dwelling having its different characteristics mapped within software, and an implementation plan that needs to be tied to the property until complete. The Passivhaus approach, although it concedes that not all properties will be capable of EnerPHit's intended level of performance, doesn't however envisage any other outcome. So within PHPP, MVHR is the only conceivable ventilation system, even though it may not be possible to reach the 5m^3/m^2.hr (at 50Pa) level of airtightness, below which, questionably, MVHR becomes worthwhile. If only PHPP could be applied to the everyday situations that are the realm of SAP, because EPCs are undoubtedly subject to the 'performance gap' between design estimates and reality. Research by the Better Building Partnership (Better Building Partnership 2012) illustrated that EPCs can in no way be relied on as a guide to energy costs, the gap being a function of the building's condition and its systems, and in particular the lifestyle of its occupants. So, decisions need to be driven by data, sensors that can feed the numbers characterising environmental conditions, occupancy patterns and the operating efficiency of systems, into the software. The mammoth task of upgrading the nation's stock demands the manipulation of big data, but since even the much-vaunted installation of smart meters has been far from successful, the political will is, as ever, clearly lacking.

The few completed EnerPHit projects have virtually turned into new-builds by time upper floors have been stripped back to make way for duct runs, ground floors have been insulated, and windows replaced with triple glazing. The question of financial viability is even more pertinent than in relation to Passivhaus new builds. The extent of insulation that EnerPHit requires is likely to necessitate bespoke window sills, and sub-frames, to accommodate the thickness of the insulation (although this additional thickness becomes less significant at each revision of the Building Regulations).

Similarly, the extra cost of triple glazing may be worthwhile on north-facing elevations where windows only lose heat. But given the energy balance throughout the heating season (the admission of solar gain compared with heat lost through double glazing is more or less evenly balanced on south elevations) so EnerPHit's blanket requirement for triple glazing is also questionable. When the additional cost of MVHR and triple glazing is compared with the actual energy savings, EnerPHit as a universal strategy is called into question. Instead, the outlay could be used for renewables, helping offset the growing power demands of household appliances and equipment.

So, the doubts that are often expressed regarding Passivhaus are equally applicable to EnerPHit. The addition of more and more insulation implies an increasing payback period with each increasing millimetre. If you add the additional costs of insulation, triple glazing and MVHR, the tens of thousands of pounds additional cost come at only a marginal saving in energy. MVHR, as ever, attracts questions because although the efficiency of the units is quoted as around 90 per cent, their gross efficiency is only 20–30 per cent; consequently, only 90 per cent of 30 per cent of the incoming air is being warmed, which generally only recovers a few degrees of heat (Marshall 2021). At times of cold weather, the pre-warming is insufficient, despite the underlying logic of Passivhaus dwellings that a heating system becomes unnecessary, so a backup electric heater battery is installed within the supply air system. In total, the payback period to achieve the standard can be in excess of 30 years, so it is not surprising that very few EnerPHit projects have been completed to date. Whether 'fabric first' can be left at a lesser specification, any money left over being used to purchase PV, or solar hot water panels, to eventually reach zero carbon, depends on factors such as the building's shape, orientation and degree of overshading. The decisions require tools and software, instead of the randomly placed panels to be seen on roofs around the country. In response, there is now EnerPHit Plus Certification (Passivhaustrust n.d.), which incorporates renewables into the list of step-by-step measures. PHPP is more than an energy modelling tool, enabling assessment of design alternatives, comfort conditions and even costs. Its further development offers our best hope so far.

The Passivhaus philosophy, relies on technology, which is regarded as having a consistent level of performance, and similar discipline is required of the building's occupants. The whole dwelling is to be maintained at a constant temperature, windows aren't to be opened in winter, other than when MVHR fans are broken, or there are no replacement filters to hand. This is one of several cultural differences between here and the engineering-driven solutions usual in Germany. Interestingly, it is claimed by at least one controls company (Atamate n.d.; BCIA n.d.) that a house with sophisticated monitoring of fabric, services and user behaviour, with controls that can switch radiators on and off if rooms are unoccupied, is capable of meeting and undercutting the Passivhaus annual heating requirement. Another alternative, enthusiastically described by Duncan Roberts, working in the Walter Segal tradition, is a design that captures solar gain in winter and eliminates cold bridges, whilst in summer employing shading and cross ventilation to reduce overheating to a few days a year. This relatively low-tech approach can, of itself, achieve significantly reduced overall energy consumption; his own self-build house runs on 118 kW/m^2/year.

> There is, however, nothing quite like being in the house on a clear but frosty winter's morning and feeling temperatures rise perceptibly as the sun reaches high enough to pour through the windows. Rather than a passive, static or remotely programmed construct, the house feels like a responsive, benign entity, attuned to the natural world it inhabits ... By obsessing as it does over the numerical performance of the fabric and components of a building there is a distinct danger that the Passivhaus approach encourages a corresponding passivity on the part of occupants. For many designers within the green movement the

most important function of a building is to enable its users to live lives that are as sustainable as possible. It is this social aspect of environmentalism that transcends all other considerations. Buildings should empower and delight their users, and enable them to become active participants in the choices that confront us all.

(Cousins 2013; Roaf 2012; Roberts 2012)

Duncan might be more sympathetic to the alternative 'Active House' approach. Although this Scandinavian (rather than German[1]) initiative is concerned with energy reduction, it takes a more inclusive view of construction and retrofit, human health and well-being are focussed on the indoor and outdoor environment. An Active House has to meet a target framework that 'balances the interaction between energy consumption, indoor climate conditions and impact on the environment.' Priority is given to the satisfaction of occupants and healthy indoor air quality through manual control of ventilation rather than automated systems. Given the diverse nature of the building stock, none of these approaches should be disregarded, or thought mutually exclusive, whether technology-driven (Passivhaus), controls systems that are responsive to and respond to the occupants, or buildings that, through sensory delight, inspire and engage their users (Active House 2020).

A similar balance is the goal of the BREEAM Refurbishment and Fit-Out Framework (BREEAM n.d.) that was first introduced as a 'domestic refurbishment standard' during the run-up to the Green Deal in 2012. Two-thirds of a project's score reflects its improvement in energy performance, assessed within SAP or EPC ratings, and the extent of demand being met by renewables. The remaining third is awarded to a variety of measures, including the rating of white goods, energy-efficient lighting, smart meters and bike storage.

A first step should be in making good the long-term vacant properties that increased by 4.5 per cent through 2019, for example, to a total of 225,845, equating to £57 billion-worth of vacant stock. Local authority and housing association ownership account for 17 per cent of the housing stock amounting to 4.5 million homes, another clear starting point for scaling-up and driving down retrofit costs. This was borne out by the 2018 IET Report *Scaling up Retrofit 2050* (IET 2018), which reiterated Brenda Boardman's view (Boardman 2005), expressed now several years ago, that the UK's emissions reduction targets will only be reached if every dwelling in the country is subject to 'whole house retrofit' but achieved at one shot rather than by incremental improvements. The main barriers to progress being those that dogged the Green Deal, a lack of demand, a paucity of consistent government policy, inadequate supply chains and associated high costs, and a lack of finance.

The challenge is now that although the decarbonisation of the grid is proceeding apace, there won't be enough electricity to go around, including powering-up electric cars, unless household heating demand is decarbonised too. So, while the Building Regulations and EnerPHit are finding their way around a gradualist approach to step-by-step retrofit, other imperatives are demanding a faster and all-at-once approach. The IET recognises that such a decisive and long-term strategy needs determined political action, but is that possible within our current laissez-faire economy where a successfully developing initiative can come to nothing at the next recession? The all-too-familiar problems are revisited in 'Scaling up Retrofit 2050': That projects so far have been too small to create economies of scale; that there are such a variety of built situations as to undermine standardisation; the lack of technical standards, tools and methodology; the indifference of consumers and the market; problems that remain intractable. Conceivably though with the application of modern methods of construction, including 3-d printing, bespoke solutions could become both feasible and attractive, but the hiatus of the last ten years hasn't encouraged innovation.

The same motivation, aligning building standards with the likely future availability of electricity to run heat pumps, underlies the 2020 *LETI Climate Emergency Design Guide* (LETI 2020). LETI in 2020 was concerned with new buildings; its proposed levels of fabric performance are comparable to or even more stringent than those of Passivhaus. But by the time of its publication, the importance of embodied energy had been realised, and the need for building construction to be part of the circular economy, every component and material to be sourced appropriately and designed for eventual reuse so that buildings become 'materials resource banks.' A whole life 'net zero' building is in these terms one that meets the criteria, not only in relation to its operational energy, but achieves complete circularity – requiring renewable energy to be used throughout, including the transportation required for its assembly, to its final disassembly. Consistent with this inclusive approach, consideration is given to the unregulated part of household electricity consumption which is the increasing number of appliances and devices. Whereas the number of households in the United Kingdom has only risen gradually since 1970, ownership of televisions and DVD players, microwaves and, particularly, mobile phone chargers has vastly increased, which, including energy for cooking, now constitutes a fifth of domestic carbon emissions. LETI wants this consumption to be metered and monitored, the aggregated date made available to a national database, and for small residential buildings to have 100 per cent of their annual energy requirement to be generated on-site. Life cycle cost analysis, and whole life carbon analysis, to be carried out for all projects, with reference to an initial brief that outlines the targets and strategies for achieving them.

LETI achieves the most comprehensive understanding to date of 'zero carbon,' a term otherwise bandied around without any clear definition. It is usually taken to mean that on-site renewables, or offsetting associated with the building, cover the annual carbon cost of regulated energy use. LETI's cradle-to-cradle approach requires consideration of whole-life costs from the procurement of materials, from a recycled source to the final reuse of the building's parts at demolition, with inclusion of all the energy uses along the way. The rigour of LETI's requirements entails close-control and monitoring, management discipline throughout, and training to increase skills. So, LETI is a utopian exercise; whether the type of loose regulation that characterises our competitive market economy can provide the collaboration needed, or whether, for example, the exchange of data that LETI requires between competitor firms can be achieved, seems highly unlikely. As with other aspects of retrofit, a successful national programme will need the state to play a leading role. LETI believes that the most onerous standards will need to be required of new construction because they won't be possible when it comes to retrofit. Brenda Boardman believed that all of the built environment will need to be brought up to Passivhaus standard with the exception of heritage properties.

In 2021, the *AECB Retrofit Standard* (AECB 2021) was published. The Association for Energy Conscious Building was an early adopter of Passivhaus principles, and their standard requires each project to be modelled in PHPP. However, they recognise that EnerPHit is likely to be too stringent, and anyway unattainable in the United Kingdom. So, their requirements, although expecting lower energy demand than that required by the Building Regulations, are much concerned with the management of risks involving moisture, flood, radon and fire. Elimination of thermal bridges is required as verified within PHPP, and the installation of MVHR or MEV is a given.

Also in 2021, LETI published its own proposals for retrofit (LETI 2021). As previously, a comprehensive approach was used in questioning 'how much is enough?' including issues of health and well-being, fuel poverty, whole life costs, cost to the national purse, and the capacity of the electricity grid. The conclusion was that the greatest benefit would be derived from deep retrofit as the norm, with up to a 70 per cent reduction in energy consumption resulting from a construction standard better than the average 2021 new build, but less extreme than Passivhaus or EnerPHit. A best practice level of retrofit that could, with the addition of heat pumps, reduce annual space heating demand down from 130 kWh/m^2 to 50 kWh/m^2. The six underlying principles were to reduce energy consumption by: Prioritising the health of the building and occupants, having a whole building retrofit plan including phased retrofit, measuring the performance, thinking big and considering embodied carbon. Recognising that dwellings are frequently constrained by their shape/form factor, heritage status or the capability of householders to withstand the extent of disruption, a range of targets from 'exemplar' to 'best practice' were defined, whether subject to modelling or being guided by LETI's indicative targets. 'Exemplars' are aligned with EnerPHit, space heating being reduced to 25 kWh/m^2/year; whilst 'best practice' relates to the AECB Retrofit Standard of 50 kWh/m^2/year, allowing a range of U-values, at least as far as walls and windows are concerned, but MVHR is a requirement in all cases. The LETI report rehearses all the familiar societal benefits of retrofit: Improved health and comfort, the employment-potential of large-scale implementation, whilst mindful of the current national heating demand for spaces and hot water, which is nearly double what it might conceivably be in 2050, if the country has by then been retrofitted, and decarbonised.

LETI's methodology is much the same as that used ten years ago by the Energy Saving Trust (EST 2008), a classification of the domestic building stock defined and modelled by house type and age of property, and distribution in terms of performance. For the lack of a national source of housing data, the archetypes are an abbreviated version of the 14,000 different types within the 2011 English Housing Survey, reduced down to just five key parameters – dwelling age, dwelling form, wall construction, window and loft insulation. The outcomes from the model have been aggregated to reach overall national reduction targets. But crucially 'how to do it' remains the perpetual sticking point. LETI's 'whole building retrofit plan' is to eliminate the now well-documented problems of piecemeal, ill-considered upgrades, guided throughout by checklists towards the completion of commissioning, and monitoring of the outcomes. Unfortunately, LETI has little concern for money and, in some cases, practicality. Addition of insulation to ground floors entails disruption to kitchen units, staircases and the residents; nowhere is the word 'decanting' mentioned in any of these standards, although that might be the highest of all project costs. The blithe assumption is made that MVHR can be accommodated in most existing situations, despite the loss of performance that follows from contorted duct routes. The case studies with which it concludes are mostly one-offs, including those very high-budget retrofits that were part of the Retrofit for the Future Competition.

LETI is a standard amongst the standards, it attempts to weave a course between the various others. But modelling isn't a guarantee of future performance. Ideally, of course, by now, there would be a national building stock database, or at least a completely retrofitted transitional town providing analysis prior to and post-retrofit, including occupancy data. The reliability of data being one of the principal issues for sourcing of materials and calculating embodied energy.

Now the pinnacle of the 'zero-carbon pyramid' the *UK Net Zero Carbon Buildings Standard* (NZC n.d.) is under development, which is to have the highest level of energy efficiency, both operational and embodied, and to finally clarify the confusion between 'zero carbon,' 'net zero' and 'carbon neutral,' pulling together all the myriad array of previous analyses, databases and standards. Perhaps this new standard will draw upon 'doughnut economics' to require that no new materials should be used, instead that all new construction should 'mine' the existing building stock for reuse of components and materials. This issue, which is rapidly climbing the agenda, has particular importance for retrofit at scale. A rolling programme for whole streets with similar house types, or a town-wide initiative for renewal coinciding with maintenance schedules, could reclaim and upcycle components and materials, a local community resource to reduce costs for residents. One-fifth of the housing stock still dates from before 1919, much of it built from brickwork with lime mortar that can be readily removed so the bricks can be reused. There is of course a flourishing circular economy for metals; scrap yards have long been in existence, as have those for architectural salvage, but inclusion of less-valuable materials will require a market for reuse to be established, either by descending to the 'rag-picker' economy of less fortunate countries, or as a result of regulation. For example, to separate and reclaim bricks from cement mortar is time-consuming and subject to many breakages, an uneconomic proposition that, if made a legal requirement, will increase project costs. The same will be true at the introduction of whole-life-costing, which amplifies the initial capital cost by inclusion of the future costs of maintenance and disposal. Not surprisingly, the very word 'regulation' has long been regarded as the enemy of free-market economics, or can we continue to hope that technology will come to the rescue?

> But the good news is that we can live perfectly well while requiring much less material. We can make lighter products, keep them for longer, and use them more intensively. We can make less scrap in production, and we can re-use components from old products.
> (Attwood and Cullen 2011; Attwood 2012)

So, we are some distance from an achievable national plan, that is fully costed for feasibility, least of all one that can achieve cross-party support and a long-term plan for its implementation, meanwhile the Construction Leadership Council estimates that the country is 400,000 short of the number of skilled workers necessary to hit Net Zero. Also, the country has less than 2 per cent of the workforce required for wide-scale increase in domestic energy-efficiency. LETI takes the view that phased refurbishment, step by step in line with maintenance schedules is no longer sufficient for the task. This is a departure from accepted wisdom, for example, the London Climate Change Partnership established in 2013 (London Climate n.d.) advised social landlords that efficiency savings were possible if 'decent homes' works, such as replacement bathrooms or flood risk reduction, were combined with climate-change adaptations. Project Calebre (University of Loughborough 2013) pointed to the implications of step-by-step measures that the performance of each successive intervention depends on the work already done.

The project compared different scenarios, such as a series of steps driven by energy considerations, as opposed to an alternative driven by affordability. For example, a boiler replaced early in the process may eventually be oversized, and operate less efficiently, once the building is better insulated. Replacement glazing and wall insulation are highly beneficial, so if carried out early, cumulative carbon emissions can be reduced. However, an all-at-once whole-house retrofit lowered cumulative emissions by 46 per cent, compared to an average of 26 per cent for staged retrofit carried out over a 25-year period. The lack of direction towards either phased or one-off whole-house retrofit is a matter of, as yet undetermined policy. Instead, we have the ranks of the converted and well-meaning providing yet more reports and standards while the world of so-called 'ethical capitalism' waits for a market solution. Meanwhile, as I write this, we are heading into the hottest weather ever recorded in this country.

In July 2020, the Institute for Public Policy Research published *All Hands to the Pump, A Home Improvement Plan for England* (IPPR 2020) rehearsing the well-worn script: 275,000 retrofit jobs possible in England by 2035, and 325,000 across the United Kingdom, whilst reducing inequality, fuel bills and improving the nation's health. By then, the rate of implementation was lagging the 2050 zero carbon goal, less than 2 per cent of the requisite annual number of heat pumps were being installed, and the number of energy-efficiency retrofits had reduced to 12 per cent of that required, due to budget cuts. Damningly, the government's existing policies were seen to be penalising the poorest households. The flagship fuel poverty programme ECO is underfunded, and regressive since it is paid for through energy bills, poorly targeted and far too slow. The cost of heat pumps and retrofitting was estimated at £10.6 billion a year up to 2030, shared between the private and public sectors, the latter making use of what at the time were very low interest rates, now a vanishing possibility. The IPRR hoped for the accelerated improvement of social housing, to create supply chains for the rest of the stock. Homeowners and private landlords would receive means-tested government grants, accompanied by low-cost loans, to meet half the cost of a heat pump. The familiar arguments in favour of variable stamp duty, and council tax skewed towards best-performing properties, were revisited, along with local authority-designated street-by-street retrofit programmes. The crux of the issue: '... developing policies to finance the upfront costs in a way that does not penalize the poorest households... such a programme needs to recognize the scale of ambition and financing that will be required and the gaps in current initiatives... the lack of scale in policymaking, the unfair distribution of costs and how, in the absence of sufficient policy, households face many barriers to funding retrofits themselves.' Raising income tax is posited as the equitable solution and a change in monetary policy towards green investment.

Given the extent of the climate crisis, the hoped-for 1.5°C limit on global temperature rise seems to be a diminishing prospect. A most pressing issue for retrofit is that we are still no closer to finding a funding formula that will make a street-by-street programme feasible. After years of inaction the Energy White Paper in December 2020 (GOV.UK. 2020), reiterated the need to take a whole house retrofit approach but contained unclear arrangements within the 'Heat and Building Strategy' that was to set a path towards the decarbonisation of heat. The Green Homes grant scheme was scrapped in March 2021 because of poor uptake; it was replaced by the Green Homes Grant Local Authority Delivery Scheme that invites councils to bid for funds to improve low-income homes. The paltry sum of £60 million was added to upgrade social housing with an EPC rating less than 'C' for the next phase of the Social Housing Decarbonisation Fund, and a commitment made to increase ECO to £1 billion from 2022. In April 2022, the Clean Heat Grant replaced the Domestic Renewable Heat Incentive (RHI), in recognition that heat pumps are expensive, and that retrofit is a necessity before their installation. So, a minimum insulation requirement is to be part of the eligibility criteria based on EPC ratings, an approach similar to the discontinued RHI. A never-ending flurry of policies and proposals, but still retrofit is the stuff of case studies and enlightened local authorities; large-scale countrywide retrofit remains a hopeful dream.

None of these approaches are completely and fully costed, there is no clear distinction between cash and non-cash aspects of retrofit. The societal benefits in terms of better health outcomes from improved housing, that children do better at school, and the employment potential, are just three of the non-cash aspects that are not included within current approaches. This is borne out by the London Climate Change Partnership's 'The Business Case: Incorporating Adaptation Measures in Retrofits' (London Climate Change Partnership 2014) that estimated the impact of exogenous aspects, ranging from the reduced impact on the NHS of respiratory illnesses, reduced heat cramps and heatstroke. The economic benefit of increased productivity due to better housing conditions was thought to be hard to quantify. If these non-cash factors were costed-in, the equation would look very different, but as with other disbenefits to the common good, such as pollution of the air by aviation, or the seas by shipping, these accrue to the public purse, not private profit.

A pioneering approach was taken by Arup when researching retrofit options for the New Barracks Estate in Salford in 2012 (Arup 2012). To complete their cost-benefit analysis, the societal values were assessed using the New Economics Foundation's Social Return on Investment tool. This allowed quantification of both qualitative and quantitative aspects, concluding that for every £1 spent, the return to the community was around £1.58. These advantages included reduced repair costs, increased training and employment, as well as reduced carbon.

The fact that within conventional accounting the numbers don't stack up is clearly demonstrated by the efforts of Bankers without Borders (Civic Square 2022). The returns on cash are poor because payback times are lengthy. Within their model, £35,000 is estimated as the retrofit cost per home, excluding the possible efficiencies of scaling-up that could reduce that to £30,000. Because the timelines are so long, some of the energy savings will be given back in the form of maintenance and replacement costs. Their assumption is that it would take 40 years to start showing a return, a purely market approach being insufficient, a combination of private and public money will be required. Although the timeline isn't suited to individuals, most of whom would be dead before the payback is achieved, it is suited to the long-term horizon of pension funds, that would need properties to be aggregated into community-sized packages. This proposal gives some of the money saved through energy improvements, back to the community as reductions in bills, and captures the rest to recompense the pension fund; it seems that pension funds could then cover 60 per cent of the capital cost. The 40 per cent gap has to be a grant; the government recoups 25 per cent through VAT and some through Corporation Tax and Income Tax. Many organisations would benefit; for example, mortgage companies would be happy as they are under regulatory pressure to improve EPC ratings, so they could be contributors. Similarly, concomitant neighbourhood improvements such as green spaces would improve the extent of runoff, and improve the lot of water companies, who would consequently be obliged to contribute. These many beneficiaries would be slicing away at the remaining 40 per cent, leaving the remainder to be within the capability of the public purse. But no money comes without strings; can we rely on the market to act in the best interest of communities? Perhaps, given that shareholders are demanding more attention to the public good, but remembering places like Bhopal, India, maybe not.

Note

1 And around Europe there are other approaches: Effinergie (France), MINERGIE (Switzerland) and Effizienzhaus Plus (Germany).

References

Active House. 2020. *Specifications for Residential Buildings*. 3rd ed. Brussels: The Active House Alliance. Available online: https://www.activehouse.info/ (accessed 14 June 2022).
AECB. 2021. *Retrofit Standard*. Available online: https://aecb.net (accessed 6 June 2022).
Arup. 2012. '*Report for Salix Homes on retrofit options for the New Barracks Estate in Salford*'.
Atamate. n.d. Available online: http://www.atamate.com/ (accessed 20 June 2022).

Attwood, J. 2012. 'Questing beast of energy policy'. *Cambridge Alumni Magazine issue 65*.

Attwood, J., and J. Cullen. 2011. *Sustainable Materials – With Both Eyes Open: Future Buildings, Vehicles, Products and Equipment – Made Efficiently and Made with Less New Material (without the Hot Air)*. Cambridge, UK: UIT.

BCIA. n.d. 'Building Controls Industry Association.' Available online: https://bcia.co.uk (accessed 18 June 2022).

Better Building Partnership. 2012. 'A Tale of Two Buildings: Are EPCs a True Indicator of Energy Efficiency?'. Available online: https://www.betterbuildingspartnership.co.uk/tale-two-buildings (accessed 12 August 2022).

Boardman, B. 2005. *Examining the Carbon Agenda via the 40 Per Cent House Scenario*. Oxford: Environmental Change Institute, University of Oxford.

BRE. 2023. Available online: https://bregroup.com/news/poor-housing-will-cost-over-135.5bn-over-the-next-30-years-without-urgent-action (accessed n.d.).

BREEAM. n.d. 'Refurbishment and Fit-Out Framework.' Available online: https://bregroup.com/products/breeam/breeam-technical-standards/breeam-refurbishment-and-fit-out/ (accessed 20 July 2022).

BSI. 2019a. *PAS 2035: Publicly Available Specification*. Available online: https://www.bsigroup.com/en-GB/standards/pas-2035-2030/ (accessed 13 July 2022).

Civic Square. 2022. 'Retrofit Reimagined: Bankers without Borders. Rufus Grantham at the Retrofit Reimagined Neighbourhood Festival'. Panel with Rebecca Lane, George Simms, Rufus Grantham, Indy Johar + Dan Hill. YouTube, 14 July. Available online: https://www.youtube.com/watch?v=jW0YmC1-p24&list=PLgIJxOKjWOpigXk97jKPuBLf5fsfGNWjb&index=7 (accessed 15 September 2022).

Cotterell, J. and A. Dadeby. (2012), *The Passivhaus Handbook: A Practical Guide to Constructing and Retrofitting Buildings for Ultra-Low-Energy Performance*. Düsseldorf: AbeBooks.

Cousins, S. 2013. 'Moisture in Buildings: Making Homes Airtight to Save Energy Could Lead to Catastrophic Moisture Build-Up'. *RIBA Journal 'Products in Practice'*. February–March edition.

EHCS. 2006. 'English House Condition Survey 2006.' Available online: https://www.data.gov.uk/dataset/f2f2fd99-cea6-41fb-ade0-1b4f22bb709b/english-house-condition-survey (accessed n.d.).

EHS. 2021. 'English Housing Survey 2020 to 21.' Available online: https://www.gov.uk/government/statistics/english-housing-survey-2020-to-2021-headline-report (accessed 10 June 2022).

EST. 2008. *Energy Efficiency and the Code for Sustainable Homes Levels 3-6, CE290-293*. London: Energy Saving Trust.

GOV.UK. 2020. 'Energy White Paper (2020) Powering Our Net Zero Future'. December. Available online: https://www.gov.uk/government/publications/energy-white-paper-powering-our-net-zero-future (accessed 6 September 2022).

Grosvenor Estates. 2019. *Heritage and Carbon: Can Historic Buildings Help Tackle Climate Change?* Available online: https://www.grosvenor.com/heritageandcarbon (accessed 10 March 2022).

IET. (2018), 'Scaling up Retrofit 2050: Why a Nationwide Programme to Upgrade the Existing Housing Stock Is the Only Way for the UK to Achieve Its Carbon Saving Goals'. London: Institution of Engineering and Technology. Available online: https://www.theiet.org/impact-society/factfiles/built-environment-factfiles/retrofit-2050/ (accessed 20 September 2022).

IPPR. 2020. 'All Hands to the Pump, A Home Improvement Plan for England', Institute for Public Policy Research. July. Available online: https://www.ippr.org/files/2020-07/all-hands-to-the-pump-july20.pdf (accessed 3 June 2022).

LETI. 2020. *Climate Emergency Design Guide*. Available online: https://www.leti.uk/cedg (accessed 6 June 2022).

LETI. 2021. *Climate Emergency Retrofit Guide*. Available online: https://www.leti.uk/retrofit (accessed 6 June 2022).
London Climate Change Partnership. n.d. Available online: https://climatelondon.org/lccp/ (accessed 16 June 2022).
London Climate Change Partnership. 2014. 'The Business Case: Incorporating Adaptation Measures in Retrofits'. Available online: https://climatelondon.org/wp-content/uploads/2014/02/The-Business-Case.pdf (accessed 3 July 2021).
Marshall, L. 2021. 'Is Passivhaus Overrated?'. *Architects Journal Specification*. July–August edition.
NHBC. 2013a. *Lessons from Germany's Passivhaus*. London: NHBC Foundation. Available online: https://www.Passivhaustrust.org.uk/news/detail/?nId=184 (accessed 6 July 2022).
NZC. n.d. 'UK Net Zero Carbon Buildings Standard'. Available online: https://www.nzcbuildings.co.uk/ (accessed 10 May 2022).
Partington, R. 2013. 'Don't Be Beaten by Technology,' *RIBA Journal, March.*
Passivhaustrust. n.d. 'EnerPHit Plus Certification'. Available online: https://www.Passivhaustrust.org.uk/news/detail/?nId=992 (accessed 10 June 2022).
Prasad, S. 2012a. "Penoyre and Prasad's Retrofit for the Future Project." *RIBA Journal*, 1 November: 70.
Roaf, S. 2012. 'Your Home – Your Choice: Pay to Run a Machine or Invest in Free Energy'. *Green Building* (4) Winter edition.
Roberts, D. 2012. 'A Response to a Recent Passivhaus Ventilation Article'. *Green Building* (56) Winter edition.
Taylor, M. 2012. *The Passivhaus Refurbishment Standard from the Passive House Institute*. London: Passivhaus Trust. Available online: https://www.Passivhaustrust.org.uk/ (accessed 10 March 2020).
'There is a Passivhaus Rush On, but We Need to Ask How it will Work'. (2008), *Architects Journal* 28 February 2008: 60.
Traynor, J. 2020. *EnerPHit: A Step-by-Step Guide to Low Energy Retrofit*. London: RIBA Publishing.
TSB. 2009a. *Retrofit for the Future Competition Brief*. London: Technology Strategy Board.
TSB. 2009b. *Evaluating Energy and Carbon Performance in the 'Retrofit for the Future' Demonstrator Projects Version 2*. London: Technology Strategy Board.
TSB. 2014. *Retrofit for the Future: Reducing Energy Use in Existing Homes. A Guide to Making Retrofit Work*. London: Technology Strategy Board.
University of Loughborough. 2013. 'CALEBRE – Consumer Appealing Low Energy Technologies for Building Retrofitting', September. Available online: https://www.cibse.org/media/cw4nmpyi/calebre-cibse-presn-v3-jul13-d-loveday.pdf.
White, C. 2011. 'Kick the Habit'. *RIBA Journal* October: *67–68.*
Young, E. 2011. *RIBA Journal*, April.

7 Pre-requisites for Retrofit

Several years on from Craig White's article, the end result will be even less easily achieved. Delivery is made impossibly difficult by the lack of coordination and skills. There are, for example, thousands of boiler installation and glazing companies, all dedicated to a single task but without the ability to take on the retrofit task as a whole. Retrofit is often spoken about as if the means are readily available, but the process will be complex, with dimensions both temporal and social, and to a large extent the technology is undeveloped. The costs relative to returns are still hard to sell in a market economy, although the rocketing price of energy is making the balance closer. Deep retrofit will not be worthwhile unless homes can be made sufficiently insulated, and airtight, that their demand for heating energy does not eliminate a large proportion of the power being generated by panels on their roofs. In many cases, it will only be possible for work to be carried out in sync with the maintenance cycle of components, so windows and doors, for example, are not being replaced before the end of their useful life. The programme will need to be coordinated with the preferences of the occupants, some of whom may have to be temporarily decanted to alternative accommodation. So, there are human and community aspects, not least because the improved homes will only save energy, if the occupants are knowledgeable about the extent to which different lifestyles can impact a dwelling's real performance.

Retrofit at the urban scale

Carried out at sufficient scale, this urban renewal will change the face of towns and cities to the greatest extent since post-war slum clearance, not only through retrofit itself but also the accompanying infill to increase density, and improvements to the surrounding environment. Yet this multi-billion-pound effort has yet to attract any significant interest, or attention, from architects. Retrofit is currently understood, at this time of climate crisis, as the work previously described as 'refurbishment,' that is transforming old industrial buildings, adapting them to use as an arts or other cultural centre. Meanwhile, the real challenge, making whole swathes of the country appropriate to the demands of the future has been mostly ignored. In the same way, Victorian architects regarded the new engineering structures of glass and steel as not worthy of their consideration. That comparison is entirely appropriate because embracing street-by-street retrofit will require a change in mindset.

Retrofit origins

For a brief period, following the economic crash of 2008, retrofit rose up the architectural agenda. Attention shifted away from shiny new buildings towards the existing building stock, new construction having been put on hold. Enthusiasm was at its height when, in 2009, the UK labour government launched a competition 'Retrofit for the Future' that sought innovative ways of improving the energy performance of the nation's housing stock. Many interesting, and innovative solutions were put forward, though it was difficult to compare their performance for the lack of a common monitoring regime, just one of the still current shortfalls. The follow-up Green Deal, a funding mechanism to enable large-scale retrofit, was unsuccessful. So, in the wake of the banking crisis, along with diminished concern for the environment generally, the initiative was shelved, never to re-emerge. In the meantime, such small-scale projects as happened often had poor performance, condensation, mould and water ingress, culminating in the energy 'upgrade' of Grenfell Tower, where the results were catastrophic. Street-by-street improvement of the UK's housing stock, some of the worst energy guzzlers in the whole of Europe, has remained a distant prospect.

The 40 per cent house

The rate of replacement is very slow so whether to retrofit, or selectively rebuild, is a critical issue. Brenda Boardman's *The 40 per cent House* (Boardman 2005) suggested in 2005 that a high proportion of the domestic stock was incapable of achieving a 60% reduction in CO_2 emissions.

The 40 per cent scenario required that all the worst houses should be replaced, demolition rates rising to four times the then current level, to 80,000 dwellings a year by 2016. Boardman's computer model assumptions were based on predicted household sizes, and the anticipated future energy mix; many more homes will be needed by 2050 due to population increase and the reducing size of households. The argument in favour of increased rates of demolition hinged on the lifetime energy expenditure of new-build compared with retaining, and retrofitting the existing stock. Even though rebuilding initially has the highest amount of embodied energy, it was calculated to be rapidly offset by the low-energy use of efficient new houses. Others, however, such as Bioregional, maintained that so much embodied energy is contained within existing housing that all should be retained, come what may. For lack of any national figures, Boardman assumed that 75 per cent of older houses, pre-1919, outside of conservation areas, would be candidates for demolition – some 3.75 million dwellings! Needless to say, this created a stir at the time despite the fact that at our current rate of replacement, the existing housing stock won't be replaced for 1400 years. The lack of progress made in improving the owner-occupied stock illustrates the difficulties that faced the Green Deal, relying as it did on carrots (financial inducements) rather than sticks (mandatory enforcement of Energy Performance Certificates). The 40 per cent House report was keen to point out that the national retrofit effort, if it was to be successful, also required change in behaviours, not only on the part of households but also the participating trades and professions.

The interest of architects in retrofit was limited and short-lived, as workloads in practice increased once again. Within the architectural community, retrofit is still assumed to be glorified maintenance, a technological question free of value judgements, outside of the philosophical debate within architecture. Most certainly, it very much depends on performance and how it is measured, a concern for the extended life of buildings, and the lifestyles within them, not just at the moment of handover and the photographs for magazines. Retrofit is then a low-key enterprise, architectural image and ambition are subordinated to efficiency and delivery, working with occupants towards long-term goals, and arriving at methods that can systematised for replication at scale. In fact, closer to the original aims of Modernism than the 'pseudo-Modernism' of recent times.

> It is ironic, yet somehow predictable, that Modernism – fruit of the economic ruin by two world wars, enemy of aristocratic privilege, champion of efficiency over sentiment – should finally with the Neo-Modernism of today, become the prestige style of the rich.
> (Benedikt 2005, 12)

Problem #2 Insulation

'Fabric first' is the overriding mantra of retrofit so insulation is also at the top of the list, given that space heating is responsible for over a half of domestic carbon emissions. This seemingly mundane consideration, of little innovative architectural interest, is like much else in the retrofit world, in need of creative development. Even loft insulation, the 'low-hanging fruit' when upgrading houses, isn't so simple. One manufacturer estimates that over 80 per cent of lofts are used for storage, likely compressing the insulation and rendering it far less effective, a survey showed that for three-quarters of respondents, their need to store in the loft was very important or essential.

For walls, an initial concern is what to use. Of course, it would preferably not be plastic. However, the fact remains that, for example, phenolic foam is nearly twice as efficient in terms of both insulation and cost as its nearest naturally occurring organic, or inorganic rival. This is an issue for thickness of construction, the length of fixings, and, in the case of walls lined with internal insulation, the reduced size of rooms. But of course, it isn't that simple. To use closed-cell vapour-impermeable plastic insulation at the inside of a solid brick wall could be highly problematic. In the case of 'hard to treat' homes, walls that don't keep out rainwater or have rising damp, would definitely militate against the use of internal insulation (IWI).

Solid masonry walls, and the other wide variety of traditional methods, aren't well served by current assessment models that don't provide accurate results about heat loss, moisture movement and air permeability. Internal insulation is a particularly sensitive problem, which would seem to suggest a 'breathable' solution as a safer alternative since much traditional construction is predicated on the absorption and subsequent evaporation of humidity. A vapour-permeable and ecologically more benign material than plastic is required, one that doesn't form a barrier to moisture from outside that would otherwise be free to evaporate. The most developed alternative, other than cork which is very expensive, is hemp, a variety of products are available, but whether the supply chain could increase production sufficiently to meet street-by-street retrofit requirements remains in doubt.

A Pro Clima (Proclima n.d.) type membrane has the manifest advantages of having variable vapour control for free movement of moisture during the summer months of high humidity, when the fabric can breathe and dry, whilst preventing the admission of condensation-forming vapour during the low-relative humidity of rooms in winter. Nonetheless, there are limits to the feasible thickness of internal insulation, until the temperature gradient across the walls almost guarantees interstitial condensation. A more sophisticated analysis than the steady state calculations provided by manufacturers, or readily completed using the psychrometric chart, is needed. These simple calculations assume that moisture movement in winter is from the inside only, taking no account of driving rain, for example. WUFI calculations provide a real-time assessment of the risks using a range of environmental data. This more sophisticated analysis is a particular requirement when trying to achieve the levels of insulation required to achieve the EnerPHit standard. Such are the risks associated with internal insulation that it is prudent to not treat more than 25 per cent of the external walls of a house in this way, if the rest can be externally insulated.

External wall insulation (EWI) is rather less risky since the temperature of the structure is kept that much higher, above dew point temperature, so the risk of condensation is reduced. Nevertheless, there have been many problems due to detailing, workmanship and application to walls that are already wet, trapping moisture inside. The familiar system used for years on the continent has a waterproof outer layer of epoxy render applied over a mesh, which is fixed through the insulation. Unlike traditional render, this synthetic material doesn't crack, eliminating the need for expansion joints. For spaces with high levels of moisture, such as wet rooms and bedrooms, normally upstairs, EWI has distinct advantages, not least because the thermal mass of the wall is still in contact with the room enabling it to moderate diurnal temperature swings. Despite these advantages, many of the building failures of recent years have been associated with EWI. The render is a wet trade, subject to both workmanship and weather, but in many cases poor detailing has been to blame, allowing rainwater to leak behind the insulation and find its way into the building. An appropriate solution for the most sheltered parts of the country, such as London, may not be the one to adopt in the west of Wales; as with most aspects of retrofit, designs need to be location-specific.

A happy compromise for cavity walls is to use external insulation upstairs, and internal insulation for the main living spaces below, particularly if they are open plan. The vulnerable external render is then out of the way of damage, and if the inner leaf of the wall is reliably dry, the use of foamed plastic internal insulation is feasible, at least economically. Even so, the rapid fall between wintertime room temperatures and outdoors, across a layer of plastic foam, pushes the dewpoint towards the room side of the outer wall, increasing the likelihood of interstitial condensation. In this case, a reliable vapour barrier is a necessity. The aluminium foil lining of foamed plastic panels reduces radiative heat exchange but is also vapour-resistant. The difficulty of providing an absolute vapour barrier has always been recognised in the distinction between a 'barrier' and a mere 'check.' The point of weakness is the jointing of the insulation panels, even small gaps form significant routes for both heat and water vapour. The adhesive aluminium foil commonly used to glue panels together is workmanship-dependent, and works only in the absence of moisture, or movement (perhaps satisfactory for the internal insulation of masonry walls with cavities but less ideal for warm roof construction or timber frame).

If a deep retrofit is also to insulate the ground floor, internal wall insulation will abut the insulation in the floor, eliminating a cold bridge at the junction. The EWI to the upper floor can be taken down to just above the ground floor windows so the floor joists are kept cosy, and because there will be an overlap with the IWI (terminating at the ground floor ceiling), the two overlap to reduce cold bridging. In this way, a continuous and imperforate 'tea cosy' is provided since the EWI also abuts the loft insulation at the roof eaves. A high proportion of the building's thermal mass will still be exposed upstairs, within bedrooms, helping to smooth the rate of temperature change throughout the night, whilst downstairs, the internal insulation ensures fast response to any demand for more heat. This technique allows for a variety of external appearances, a mix of original brickwork and over-cladding that can provide a more sensitive approach than the unrelenting application of EWI. If the whole of the house was to receive EWI, its external elevation would completely change, one of the perceived problems of street-by-street retrofit. Also, if EWI comes down to DPC level, to eradicate the cold bridge at the ground, the insulation should extend down to the foundations, with all the difficulty of removing paving and cutting around service entries. At the moment, even this everyday situation lacks the component necessary to easily make the junction between the below-grade insulation and the metal track supporting the bottom of the EWI.

Although rigid board insulation has been the preserve of foamed plastic, the familiar alternative, Rockwool, has been in batt form the usual choice for insulation of roofs, the easiest achieved of retrofit's energy improvements. But also, for the over-cladding of taller buildings, in the wake of Grenfell, rockwool would seem to be the reliable alternative. The introduction of dual-density slab (Rockwool n.d.) with an outer dense layer into which secure fixings can be made, and an internal insulative core, is intended for that market. To achieve the same energy efficiency, rockwool slabs have to be that much thicker, implying longer and more substantial fixings which conduct heat.

When the retrofit programme seemed on the verge of take-off, back in 2010, a variety of new solutions to the insulation dilemma were mooted. For example, there was the STO-solar system (STO 2012), an external insulation that employed the same principle as the fur of a polar bear, which has bristles that are hollow, and transmit warming sunlight, through to the animal's absorbent black skin. In the case of STO-solar insulation, panels of bundled hollow plastic filaments were installed over the black-painted surface of the outer wall, and then waterproofed with a translucent external render. The system was designed for winter efficiency when low-level sunlight, providing heat, was closest to the right-angled junction of filaments and wall.

Dynamic Insulation, whereby air is drawn into the house through the insulation and pre-warmed by the heat that would otherwise be escaping, is a concept with a long history, particularly in Scotland. In 2012, it was used for the construction of a demonstration house in Dunfermline, where monitoring equipment recorded the system's success in achieving a negative U-value for the walls. Other projects have been less successful, undermined by the force of the wind; on the leeward side of a dynamically insulated building, air may be drawn out through the walls rather than in, depending on the fan power employed.

Suspended timber floors present a particular problem since the floorboards, given that the void below is at ambient temperature, have the same insulation value as a single-glazed window. Insulating above an existing floor creates problems at doors, the staircase and kitchen units, but having to lift all the boards and insulate between the joists is also highly disruptive. A useful development is the use of a robot (Q-bot n.d.) that travels through the void, and sprays polyurethane foam onto the underside of the existing floor. Since the millions of suspended ground floors in the country are in some of the most 'hard to heat' homes (albeit that foam insulation is not ideal), this method of minimal intrusion, the complete installation is said to take one or two days, is a precursor of other AI solutions.

A project that won many awards was WHISCERS (WHISCERS 2012), a method for internal insulation pre-manufactured in kit form. Initial on-site measurements were made by laser survey, and the data was transmitted to automated CNC cutting equipment for production of the insulation panels, each coded for accurate jointing together on site. The aim was for the minimum of disruption, the occupants remaining in residence. These innovative methods and materials, particularly where production lines need financing, are vulnerable to the ebb and flow of markets. The slowdown in large-scale retrofit schemes in the intervening years has put paid to many hopeful ideas, in the same way that proposals for off-site construction have foundered in the past. Back in 2012, an industry survey commissioned by the BRE (More Innovation 2012) found a pressing need for more research into offsite construction for retrofit, and new ways of insulating existing buildings. Innovative methods of retrofit, that can be carried out with occupants in place, need to be found, as well as improved insulation solutions for solid-walled houses, using natural materials with low carbon content.

Of course, the bottom line is how much insulation? Pragmatic advice, considering the thermal and hygrothermal stress placed on traditional structures subjected to a different insulation regime than they have been unaccustomed to, can be 'moderation in all things.' The LETI *Climate Emergency Retrofit Guide* includes in Appendix G (LETI 2020) analysis of the cost effectiveness of levels of insulation applied to a solid wall, the inflection point happens around 10 centimetres, in terms of cost per watt, equivalent to a U-value of 0.3 W/m^2K, which they note 'may not necessarily equate with best practice.' To achieve the last 20 per cent of heat loss saved will require ever-increasing thickness, the law of diminishing returns. LETI recommends that, on balance, an 80 per cent reduction in heat loss would be appropriate for a poorly insulated construction, requiring a greater thickness of 15 centimetres, depending on the extent of the building's existing insulation. In an assessment of the Passivhaus retrofit of a solid-wall masonry building in Germany, it was found that increasingly marginal gains in heat retention were afforded by super-insulation, due to unavoidable thermal bridges.

On the other hand, the case can be made that thicker insulation eventually pays for itself, in previous times of very low interest rates, thicker insulation represented a good investment over time, particularly when factoring inflation into the picture. As energy prices rise, inexorably the validity of this argument increases, and the validity of simple payback times decreases. The feasible extent of insulation in retrofit situations is highly dependent on the existing fabric, but at the very least, EnerPHit provides an ambitious level representing the gold standard. But isn't EnerPHit's intention to eliminate the need for a conventional heating system itself dubious, if there is still the need for backup heat provided in the most expensive way possible – in the form of a heater coil within the ventilation system? In a warming climate with shorter heating seasons, the flexibility afforded by a pullover, and rooms at different temperatures, within adaptive comfort limits, has much to recommend it. Parity Project's study in Merton (Parity Projects 2009) illustrated that large savings can be achieved by reducing levels of heat, it was estimated that turning the thermostat down from 24°C to 18°C would produce cost and carbon savings of a third, even a reduction from 20°C to 18°C resulted in over a 10 per cent reduction. Biomimicking nature's trick, providing some creatures with a winter coat that they shed come the spring, might be the guide for future technologies. A primitive version being American storm windows that are installed or taken off according to the weather, worthy of development since it has been found in Scotland that secondary glazing reduces heat loss to an extent comparable with more expensive, replacement double glazing (Baker 2011).

Indeed, similar arguments can be had about windows generally. Developments in technology have tended towards more layers of glass, coatings, and fill with inert gas, argon, xenon or even krypton. To maintain its rigorous insulation requirement, the Passivhaus credo is for all windows in both new and retrofitted homes to be triple-glazed. For south-facing windows though, even in the United Kingdom, over the course of the heating season the heat gains (g-value) balance with heat losses (U-value). So, depending on the shape, the form factor, and orientation of the house there can be a considerable saving in first cost, and embodied energy, by simply installing double glazing, assuming best properties for the glazing and frame. The Passivhaus Institute isn't however solely concerned with energy, but also the comfort conditions that are a function of downdraughts, and the radiant asymmetry of windows compared with their surrounding areas of wall. Fortunately, contemporary double glazing with low-e and argon-fill significantly increases energy performance above that of older, simpler units so the temperature of its inner surface is increased too, limiting downdraughts (though maybe not sufficiently to reduce asymmetry sufficiently to be within the 'comfort zone'). Consequently, Passivhaus requires triple glazing throughout, but are downdraughts sufficient reason, in this land of woolly slippers (as opposed to the cold winters of Germany) a good enough reason to exclude the warmth of winter sunshine? It's a curious anomaly for one of the founding principles of passive design, the best utilisation of the sun, to no longer apply. So once again, the appropriate specification is dependent on each circumstance, at present there is a shortage of simple ready-reckoner tools that can make the choices evident for each prospective retrofit, even more so now that the embodied energy of additional layers of glass needs to be factored in.

Petrochemicals are in retreat as the oil economy wanes and the need for more ingenious alternatives becomes ever more pressing. The hundreds of varieties of polymer, each with their own individual characteristics, have found their way into every aspect of contemporary life – where are the alternatives? In the case of insulation, there is one that beguiles – aerogel – far more efficient than any other insulating material and consequently able to be used in much reduced thicknesses. It was employed in several of the Retrofit for the Future entries, some years ago, but only for detailed situations, such as around bay windows, because of its cost. Because aerogel is mostly air, it is very fragile so panels couldn't be cut to fit on-site; instead, precise components had to be pre-manufactured. On the one hand, aerogel has the advantage of incombustibility and allowing moisture movement, but it does slump over time. There have been developments, aerogel adhesive tapes, for example, but this material, once so costly that it only found its applications in outer space, is crying out for urgent implementation. If the state-funded Manhattan Project, required to rescue the world from badly performing buildings, had been instituted years ago rather than relying on market mechanisms, aerogel would be the universal solution. As Dr Peter Tsou of NASA's Jet Propulsion Laboratory put it:

> you could take a two- or three-bedroom house, insulate it with aerogel, and you could heat the house with a candle. But eventually the house would become too hot.

If there ever was a technology we need now, this is it.

References

Baker, P. 2011. *Improving the Thermal Performance of Traditional Windows*. Centre for Research on Indoor Climate and Health, School of Engineering and the Built Environment, Glasgow Caledonian University. Available online: http://energy.pjb.co.uk/improving-the-thermal-performance-of-traditional-windows/ (accessed 8 August 2022).

Benedikt, M. 2005. "Less for Less Yet: On Architecture's Value(s) in the Marketplace." In *Commodification and Spectacle in Architecture,* edited by W.S. Saunders, Minneapolis, MN: University of Minnesota Press.

Boardman, B. 2005. *Examining the Carbon Agenda via the 40 Per Cent House Scenario*, Oxford: Environmental Change Institute, University of Oxford.

LETI. 2020. *Climate Emergency Design Guide*. Available online: https://www.leti.uk/cedg (accessed 6 June 2022).

'More Innovation Required in MMC and Renewables Say Construction Professionals.' 2012. *Building for Change, 16 August.*

Parity Projects. 2009. *Carbon Assessment of Domestic Housing in London Borough of Merton*, version 1.0.

Proclima. n.d. Available online: https://proclima.com/ (accessed 14 May 2022).

Rockwool. n.d.. Available online: https://www.rockwool.com/uk/products-and-applications/product-overview/cladding-solutions/ewi-slab/ (accessed 16 May 2022).

Q-bot. n.d. Available online: https://q-bot.co/ (accessed 25 May 2022).

STO. 2012. 'STO-solar system: STO Solar Heating System Pads over to the UK'. *BRE Building for Change 2012.*

WHISCERS. 2012. 'Internal Insulation of Linden Lea and Wenlock Edge' Dorking, Surrey. *BRE Building for Change 2012.*

8 Drivers for Change

In the United Kingdom, the motivations for radical demand reduction are a consequence of other factors as well as climate change. Peak oil was probably passed around the millennium; the country is increasingly reliant on fuel imports with the attendant risks of interrupted supply and huge fluctuations in price. These pressures are finding their way into policy, for example, the *UK Low Carbon Transition Plan 2009* (GOV.UK. 2009) intended that 30 per cent of electricity would be derived from renewable sources by 2020 and that by then every home in the country would be equipped with a smart meter. Between 2008 and 2013 Project Calebre (University of Loughborough 2013) (Consumer Appealing Low Energy Technologies for Building Retrofitting) involved six universities in studying a range of technical options that might make an impact on future retrofit, including MVHR, novel methods of moisture control and vacuum glazing, one of the few efforts at evaluation independent of the claims of manufacturers.

The *Warmer Homes, Greener Homes: A Strategy for Household Energy Management* from 2010 (GOV.UK. 2010) was to reduce energy demand by implementing measures particular to the characteristics of individual properties and achieve 'whole house' solutions that are more radical than the 'low-hanging fruit' that have largely been picked by now, such as loft insulation. So, its ambitious goals were that by 2015 all lofts and cavity walls would have been insulated; by 2020, seven million upgrades would have been completed; by 2030 all cost-effective retrofits were to have been implemented, and by 2050 the carbon emissions from buildings would be as close to zero as possible. Policy documents come and go, but targets are never met, despite their grandiloquent titles, the latest is the *Net Zero Strategy: Build Back Greener* (GOV.UK. 2021).

The trajectory towards current requirements can be traced back to the work of Brenda Boardman (*40 per cent House, Home Truths, Achieving Zero*). The *40 per cent House*, written in 2005 (Boardman 2005), proposed the measures necessary to achieve the easier 60 per cent reduction that was the UK's goal at that time.

Even this more modest ambition required that all existing houses would get loft insulation and high-performance windows because the majority of households' energy

DOI: 10.4324/9781003564997-11

does, of course, go into the heating of rooms, and the heating of water. In addition, two low- and zero-carbon technologies would be needed per house, for example, combined heat and power provided on a community basis or heat pumps, biomass, photovoltaics or other renewables. The optimistic assumption within her model was that future efficiencies would enable the electricity consumption due to lights and appliances to be halved by 2050. But the likelihood of air-conditioning being installed to counter summertime overheating as the climate warms had to be discounted, in the hope that countermeasures such as shading and night-cooling would suffice.

About two-thirds of Boardman's anticipated carbon savings were due to demand reduction – insulation etc. – and one-third as a result of energy generation resulting from the installation of on-site renewables. Despite her assertion that Western homes don't need ever-increasing standards of living, allowance was made for an average additional requirement for warmth and hot water, to ward off any charges of overly green enthusiasm. The challenge within the '40 per cent house' scenario was to reduce the average space heating used by retrofitted houses to 9000 kWh, and new-builds to 2000.

Older solid wall 'hard to treat' houses are a particular problem, not least because, in 1996, 38 per cent of those in fuel poverty lived in houses with the poorest SAP ratings, many of which are pre-1919 terraced houses. Boardman's ambition for the housing stock to become a net generator of electricity by 2050 was to be facilitated by the widespread installation of photovoltaics, mostly once they were expected to reduce in price after 2030. Because CHP from fuel cells is best suited to houses with high heat demands, those dwellings only able to be marginally improved would succeed in becoming low-carbon houses because they would be exporting electricity to the grid, an interesting departure from the usual 'fabric first' approach. Growth in demand for hot water was to be offset by solar thermal panels on roofs, sized to enable the space heating system to be turned off in summer. Whether or not installation of these expensive renewables was justified for retrofit of what otherwise would be the least improvable houses was a matter for 'further debate.'

Problem #3 Overheating

Retrofit architects have to have not only a grasp of technicalities but a willingness to engage with householders and communities to ensure that energy benefits aren't just at handover but are long-lasting. A vast complication, and one that has scarcely been addressed, is that retrofit measures directed at reducing heat consumption in winter will be of diminishing value in a warming climate. Not only will winters effectively become shorter but summertime overheating will be a growing problem, to retrofit with inflexible technologies may prove a short-sighted solution. A grasp of the complexities of this problem and the skills required for street-by-street large-scale retrofit are prerequisites for the future role of architects. But given the unknowability of the future, deriving as yet undeveloped strategies will need the involvement of inventive imaginations.

There is no universally accepted limit for summertime comfort temperatures; less research has been done in dwellings than for other building types. Most of the studies on thermal comfort, including the foundational work by Ole Fanger (Fanger 1982), were carried out in mechanically conditioned buildings such as office blocks, where occupants have little control of their environment. Fanger himself agreed that in 'free-running' buildings, which are typical of the majority of UK housing stock, an adaptive comfort approach would be more applicable. Human comfort has shown to be achievable across a wider range of temperatures, up to an additional two degrees of outdoor temperature, if appropriate behaviour, taking cool showers, drinking cold drinks, adjusting clothing etc. are factored into the equation. Over time, people become acclimatised to warmer conditions (but perhaps not at the rate of global warming), but for that reason, it is hot temperatures, when occurring at the beginning of the summer, that are most dangerous to vulnerable groups and the elderly.

The overheating criterion within Passivhaus is for an internal temperature of over 25°C to not be exceeded for more than 10 per cent of the year. By comparison, CIBSE (CIBSE 1999) using a similar benchmarking approach, originally advocated no more than 1 per cent of annual occupied hours to be at a temperature more than 28°C in living rooms and 26°C in bedrooms. But CIBSE's later 'Climate Change and the Indoor Environment: Impacts and Adaptation' (CIBSE 2005), resulting from detailed studies of climate change impacts, concluded that the bedroom criterion should be modified to 1 per cent over 25°C. Bedrooms are a particular problem since after the stress suffered by occupants during a hot day, hot nights result in loss of sleep, which is detrimental to health and performance at work. Traditional buildings, in hot climates, have sleeping spaces that vary throughout the year, whether up on a flat roof when the climate is equable, or in a cool basement to escape the heat. Flexibility in the use of rooms should be a briefing requirement for retrofit since temperatures in houses can be considerably higher on the upper floor at night, thermal mass and shading to be positioned accordingly. The thermal gradient in two-storey houses and multi-storey flats militates against uniformity and in favour of flexibility.

July 2022 saw the first ever 'level 4 heat alert,' 4500 heat-related deaths are thought to have occurred that year (the estimated future cost of heatwaves to the United Kingdom could be £60 billion annually). Our rapidly warming summers have required introduction of a new section of the Building Regulations Part O (2022). Overheating is a problem in recently constructed buildings insulated to current requirements, rather than older buildings, particularly those with massive masonry walls; a cool old church is always a welcome refuge on a summer's day. The most at-risk are detached houses, which have a lot of exposed external surface. Also, top-floor flats beneath hot roofs, subject to their neighbours' rising heat, and likely to not have cross ventilation. It is a general assumption that heating loads in winter will for many years be the principal energy cost, but the future climate is of course an unknown, despite the attempts within the Prometheus programme (University of Exeter n.d.), for example, to model future outcomes. The risk is that if summertime overheating becomes unbearable, mechanical cooling will become widespread, offsetting the efforts put into retrofit. Thermal mass, as in the church interior, is valuable if it can be cooled at night to help offset daytime temperatures, but this also may be a diminishing asset if summer nights get ever warmer, particularly because tolerance to heat when trying to sleep is that much less than when up and around. A technology that achieves the heat retention of massive construction, but without its weight, is the variety of phase-change materials (PCM) based on organic compounds or inorganic salts. A rise in temperature causes wax beads embedded in a plasterboard panel, for example, to undergo phase change from solid to liquid, with the absorption of heat.

The largest national study so far, conducted by the BRE and Loughborough University, investigated 2019's indoor summer conditions in 750 English homes (University of Loughborough 2019). Pro-rata the whole housing stock, 19 per cent of bedrooms at night and 15 per cent of living rooms during daylight hours, a particular problem in flats, could be assumed to be overheating. External and cavity insulation added no additional risk; in fact, they help keep the summer's heat outdoors. Internal insulation added a small increment, but very little compared with the heat load exerted by windows. Flats with large windows, and rooms ventilated from one side only, are a particular problem as are lofts in houses. Solar panels provide shade as does painting flat roofs white (a legal requirement in California). Attention was drawn to the risk to life posed by the problem, for vulnerable groups, the elderly, babies and those living in deprived areas. The lack of security that results from windows being left open overnight limits the possibility of night-time cooling in many locations. A joined-up approach is anticipated within the forthcoming 'Heat and Buildings Strategy.'

PHPP runs on historic climate data so for many years it has been maintained that Passivhaus dwellings do not overheat. This assertion was on the grounds that a heavily insulated airtight building functions like a vacuum flask so internal conditions are maintained irrespective of those outside. The difference is, of course, that unlike a vacuum flask a house has windows. Consequently, Passivhaus designs are now subject to an additional calculation routine, PHT, to assess overheating risk, or CIBSE TM59 to show compliance with Part O. The rate of air-change of an MVHR system in summer override is nowhere near the rate that will induce a cooling effect on the skin during hot summer's days. Therefore, Passivhaus windows have to be openable, cross ventilation is advantageous but for an east-west, ideal passive solar orientation, most of the openable windows will be on the hot sunny side of the house so hot air, once admitted, will find its way upstairs to the bedrooms.

The Passivhaus recommendation is that fixed shading above windows should project from the face of the building to subtend a 60-degree angle with the wall, so low-level solar gain is admitted in winter, but the sun when high in the sky is obstructed. Unfortunately, it is difficult to keep low-level morning and afternoon sun from the glass on a south façade, whilst much of the high-level midday sunlight will anyway be reflected outwards at the acute angle it makes with the window. Internal blinds are less effective, though solar reflective blinds are available (but cannot be used to demonstrate compliance with Part O). Because the wavelength of the sun's rays changes as it passes through glass resulting in greenhouse effect, it is heat that won't readily escape through a heavily insulated envelope. The method traditional in Southern Europe, hinged shutters or roller shutters, needs the occupants to operate them, but unfortunately in the United Kingdom the limitation is that windows nearly always open outwards. Nonetheless, on the hottest of days when opening windows would not be advantageous, closed windows, closed external shutters and a darkened interior provide the only respite.

All this is reflected in the modelling study carried out in 2015 by the Encraft engineering consultancy (Encraft 2015) that looked at different Passivhaus buildings across a range of climate forecasts. As regulations grow more stringent, the outcomes become applicable to new homes generally. The chosen scenarios anticipated either medium or high emissions in the year 2080, and the Passivhaus standard recommendation that 25°C shouldn't be exceeded for more than 10 per cent per year. Two different property types were investigated, three-bedroom houses and three-bed maisonettes in Glasgow, Leicester and Portsmouth. A variety of cooling measures were modelled but in the short term much the most effective was external operable blinds, though by 2080, the distinction between this and other measures such as opening windows, overhangs or internal blinds was less marked. In fact, throughout all future scenarios the efficacy of overhangs on south-facing windows, and internal blinds, was much the same. Internal blinds do however require knowledgeable operation by the occupants so not surprisingly the change to the regulations has erred on the side of caution, external overhangs will become the order of the day. Overheating was concluded as requiring a design response before 2050, preferably as the need arises rather than incurring maintenance costs in the short term, a case for staged retrofit.

What is consistent in the approach of both the Passivhaus Institute and Part O of the regulations is to discount any engagement with the occupants or to recognise the benefits of adaptive cooling. Although indoor air quality isn't readily assessed by human perception, thermal comfort is, so conceivably occupants might open windows, draw blinds, take cold drinks, sleep on the cooler side of the house, or use ceiling fans. Instead of these everyday measures, as a last resort, Part O envisages mechanical cooling, yet another technological assertion – that compared with human responses those of machines are relatively infallible. True enough, Part O allows other methods of compliance, but the regulations always represent the lowest common denominator, so we can expect to see fixed shading on all new housing and retrofits too, irrespective of location. Surely a more ecological approach would be to plant more deciduous trees that provide shade to south facades in summer whilst letting the sun shine through in winter and encourage more soft landscaping that conserves moisture to cool the surrounding air and buildings. Ground floor windows, for example, can be effectively shaded by vegetation that may die down in winter, but if chosen for rapid growth in the spring, can prevent the sun from reaching the windows by the time the warmest months arrive. A more technical approach is taken by Glass-Xcrystal (Glass-Xcrystal n.d.), a glazing system that provides both shading and thermal mass, its triple-glazed thickness includes a layer of salt hydrate PCM providing thermal mass, and prismatic louvres that admit winter sun, and block it during the summer. As with the approach to net zero generally, there is an urgent need for innovations to be tested, verified, rapidly endorsed and manufactured at scale to reduce costs.

In summary, we live in a country with a number of different climates and micro-climates of uncertain future with traditional buildings which may perform better in summer, even though they are urgently in need of better energy performance, and newer dwellings that are already inclined to overheat. As with other aspects of retrofit, flexible technologies (perhaps PCM) are urgently required, but the long-delayed mass retrofit of the country is hampered by a lack of adequate methods, components and bright ideas.

References

Boardman, B. 2005. *Examining the Carbon Agenda via the 40 Per Cent House Scenario.* Oxford: Environmental Change Institute, University of Oxford.

CIBSE. 1999. *Environmental Design, Guide A.* 6th edition. London: CIBSE.

CIBSE. 2005. *Climate Change and the Indoor Environment: Impacts and Adaptation.* London: CIBSE. Available online: https://www.cibse.org/knowledge-research/knowledge-portal/tm36-archived-climate-change-and-the-indoor-environment-2005 (accessed 15August 2022).

Encraft. 2015. 'Technical Insight: Strategies for Mitigating the Risk of Overheating in Current and Future Climate Scenarios'. Available online: https://www.passivhaustrust.org.uk/UserFiles/File/Technical%20Insight%20-%20December%202015%20-%20Mitigating%20Oveheating%20Risk%20in%20Future%20Climates.pdf

Fanger, P.O. 1982. *Thermal Comfort. Analysis and Applications in Environmental Engineering*, Malabar, FL: Krieger.

Glass-Xcrystal. n.d. Available online: https://www.glassx.ch/en/products/glassx-crystal/ (accessed 10 May 2022).

GOV.UK. 2009. *UK Low Carbon Transition Plan*, Chapter 4: Transforming Our Homes and Communities, National Strategy for Climate and Energy. Available online: https://assets.publishing.service.gov.uk/government/uploads/system/uploads/attachment_data/file/228752/9780108508394.pdf (accessed 10 June 2022).

GOV.UK. 2010. 'Warmer Homes, Greener Homes: A Strategy for Household Energy Management'. Available online: https://www.thenbs.com/PublicationIndex/documents/details?Pub=DECC&DocID=293709 (accessed 12 August 2022).

GOV.UK. 2021. 'Net Zero Strategy: Build Back Greener'. Available online: https://assets.publishing.service.gov.uk/government/uploads/system/uploads/attachment_data/file/1033990/net-zero-strategy-beis.pdf (accessed 12 August 2022).

University of Exeter. n.d. 'Prometheus Project Weather Files'. https://engineering.exeter.ac.uk/research/cee/research/prometheus/downloads/ (accessed 20 September 2022).

University of Loughborough. 2013. 'CALEBRE – Consumer Appealing Low Energy Technologies for Building Retrofitting', September. Available online: https://www.cibse.org/media/cw4nmpyi/calebre-cibse-presn-v3-jul13-d-loveday.pdf (accessed 6 March 2022).

University of Loughborough. 2019. 'BRE and Loughborough University investigated 2019's indoor summer conditions in 750 English homes.' Available online: https://www.lboro.ac.uk/news-events/news/2021/july/over-4.6-million-homes-experience-overheating/ (accessed 10 May 2022).

9 Home Truths

By 2008, after the publication of *Home Truths* (Boardman 2007), the extent of the problem and the government's commitment to solving it, had been ramped up. Within the Climate Change Act, the target had risen to 80 per cent carbon reduction by 2050. The context is now familiar, peak oil had been, or was about to be reached, with consequent worries about energy security and the pollution caused by the burning of fossil fuel. The measures proposed were to maintain global temperature rise to be no more than 2°C, but it was clear, even then, that by 2020 the 18 per cent carbon reduction target would be difficult to achieve given that, for example, 17 million homes had cavity walls but half of them were still unfilled. The number of households living in fuel poverty had doubled to 4 million since 2002, and it was predicted that the country might need to accommodate a quarter more households by 2050, implying a 25 per cent increase in domestic energy consumption.

Just before the financial crisis, in 2008, the consultation on the Government's 'Heat and Energy Saving Strategy' (GOV.UK. 2008) was launched. An impressive array of initiatives was intended, some of which came to fruition such as Community Empowerment Networks, others including the Community Energy Savings Programme that ran until 2012, were short-lived, the proposed Heat Markets Forum never happened. Underlying the proposals was a clear statement of guiding principles: urgency, enabling people to make changes, fairness whatever one's income, and better regulation. Amongst the targets that have never been pursued was that every home in the country should, by 2030, have an established 'whole house' retrofit plan. Identified as a 'key plank' of the strategy, merely undertaking simple and cheap measures such as loft insulation, would give way to a comprehensive approach based on an energy audit, leading to adoption of the most effective installation sequence.

Boardman anticipated the extent of the challenge, but also the benefits if homes achieved warmer and healthier conditions, if fuel poverty was eliminated, and given the employment potential of low-carbon retrofit. The distinction between energy and carbon is vital. Although on average, 65 per cent of energy used is for space heating, and 22 per cent for hot water, the remaining 13 per cent for lights and appliances is important in terms of carbon because electricity is a lot more polluting than gas. We are starting from a low base; the energy efficiency of houses in the United Kingdom is poor, we use twice the amount of energy for heating as

DOI: 10.4324/9781003564997-12

compared with Scandinavia. Here the heritage lobby is so strong, and the replacement rate of the housing stock is so small, that 25 million out of 25.8 million houses will still be around in 2050, by which time their rate of heat loss will need to have been reduced by at least a half. Retrofit is a big challenge for energy policy.

Meanwhile, although in recent years improvements to existing houses have increased their energy efficiency, they haven't reduced energy demand. Instead, householders have been enjoying the benefits of extra comfort rather than using less energy. Because of the large number of unimproved homes, for each £3 spent on heating, £1 is wasted. The trajectory to 2050 was mapped in terms of Energy Performance Certificates, which use the Standard Assessment Procedure (SAP) as their basis. The average SAP in 2005 was 48 points, an improvement upon 42 points from 1996. The EPC bands from red for 'G', to green for 'A', are proportional in width to energy use, so the diminishing widths of the bands represent increasing SAP ratings. It gets progressively easier to progress from one band to the next. The largest jump in energy, carbon and cost savings is when moving from bands G to F. At that time, most UK houses fell between bands D and E, very few were in the higher bands A and B.

This was borne out by the *English House Conditions Survey* published in 2006 (GOV.UK. 2006). Across the housing stock as a whole, the largest group occupied band E, a result of the largest proportion of the private sector occupying that band. The much smaller amount of social housing did rather better, in the case of both local authority and housing associations band D was occupied by the largest number of dwellings.

The policy that *Home Truths* proposed is that by 2050, the housing stock, although 85–90 per cent of it will be as existing today, will be insulated to a good standard and generating some, or perhaps all, of its own electricity and heat. The energy efficiency of these existing dwellings will have been transformed, in some cases to a SAP rating of 80, but in all cases to a minimum of 50, which was the national average at that time. Homes were predicted to become net exporters of energy from 2040 onwards, without the use of oil or coal; gas, responsible for 65 per cent of the energy used in 1998, would be down to 50 per cent in 2050. The value of the housing stock would have been preserved whilst providing better warmth and comfort. A picture is painted of an '80 per cent house' that is fully insulated (all solid walls having been insulated), double or triple glazed, with at least one solar technology on the roof. An internal temperature of 21°C being achieved, whether or not the building is in use, and at night-time, the health effects of poor housing to be eliminated; fewer radiators would be needed, wood stoves being used occasionally in winter to make a cosy atmosphere. The occupants were to be enlisted as consumer watchdogs, ensuring that they would get the expected savings from their new items of equipment.

40 per cent house to achieving zero

As climate change predictions were being ratcheted up (now to a feared increase of 4°C), in 2012 Brenda Boardman produced a further series of recommendations in *Achieving Zero – Delivering Future Friendly Buildings* (Boardman 2012). The emphasis here was on occupants, rather than physical measures, so as to better align property values with their energy efficiency. The proposal was made that mandatory minimum standards should be legally enforced; consequently, by 2050, the average UK house would be at the top of EPC band A, using zero net energy technology household electricity use would be offset by on-site renewables. The beginnings of this approach were on their way with the proposed legal requirement for all band D to G properties to be improved; otherwise, they wouldn't be able to be rented; a good intention that has since been abandoned.

Problem #4 Ventilation

Ventilation heat loss is significant; in older homes, it constitutes about 20 per cent of the total, but in newer dwellings with better insulation, the percentage rises to around 35 per cent. Despite the energy implications, not to mention implications for human health, ventilation has not been at the forefront of retrofit concerns. Now, however, ventilation has stepped out of the shadows. Whereas, back in the days of the Green Deal, ventilation measures were not even eligible for funding, subsequent problems, particularly those involving internal insulation, have sent ventilation issues to the top of the agenda. It is now the first question to be answered when making an assessment within PAS 2035. In the same way that the energy consumption profile of every dwelling is different, according to its construction and occupancy, the same is true for ventilation, it is the occupants and their machines that generate moisture and CO_2.

> 'Bathrooms these days are ventilated with electric fans not to remove noxious odours but to evacuate the even more noxious water vapour. Houses seem to deteriorate from the bathrooms out...'
> (Brand 1994, 114)

The current wisdom is that airtight dwellings, in accordance with the mantra 'seal tight, ventilate right' should be mechanically ventilated. But those less well-sealed (above 5 m^3/m^2.hr when pressure tested at 50Pa) until the 2021 revision of the Building Regulations, could have background natural ventilation supplied by trickle vents and extracted through passive stacks. Or alternatively, intermittent fans (some of which, such as the Inventers (Inventer n.d.) system, can now achieve levels of heat reclaim similar to mechanical ventilation heat reclaim, MVHR). Intermittent fans in wet rooms are now, within the updated regulations, the only alternative available for less well-sealed houses, likely to be the majority of older retrofitted homes (though this technology seems likely to be phased out by the forthcoming

Future Homes Standard, leaving ducted systems as the only alternative). Intermittent ventilation is regarded by the Passive House Institute as just about meeting the energy conservation law in Germany, and adequate for the refurbishment of existing homes. But, because the air supply may be too cold for comfort in winter, the inlets need to be situated above radiators. Curiously, when the fans aren't in operation, this system relies on the adventitious ventilation provided by infiltration and trickle vents, maybe excessive from an energy point of view; or depending on wind conditions, insufficient for health. In other words, precisely the objections made to passive stack. Now passive stack has been shuffled off back into the shadows, and fans are the only promoted option. Demand control for regulated operation, using humidity-activated vents at the inlet of passive stacks, has seemingly been disregarded despite their impressive credentials. But even in southern England's relatively benign winter months, there are many examples of trickle vents being closed or blocked due to draughts. Ventilation has only recently been highlighted as a critical issue, but one where there are many unanswered questions. This latest revision to the regulations appears, as ever, to have been written cognisant of the desire of industry to sell fans and ducts.

So, it is a strange anomaly of current low-carbon thinking that passive energy and natural ventilation should be espoused, and increasingly practised, in relation to non-domestic buildings, (Lerpiniere 2022) and even for some challenging building types such as theatres and television studios, whereas future housing is being directed towards sealed construction and mechanical ventilation. But this logic isn't uncontested, even though it is enshrined within the Building Regulations. We live within a currently temperate climate, on a windy island, surrounded by relatively warm sea. So, in winter, there is enough temperature difference between inside and outside to drive passive stack ventilation assisted by the wind, and perhaps with help from in-line fans at the top of the stacks that just come on as necessary. In other words, hybrid ventilation, as widely accepted for non-domestic use. Certainly, in extremely airtight dwellings the ventilation system also needs to reclaim heat, and from that point of view, MVHR is the only competitor, particularly as its efficiency has increased over the years. Nowadays, the operation of MVHR is thought cost-effective even if the external temperature is as high as 17°C, and internally 20°C, the Passivhaus norm.

The reasoning that passive stack is no longer an approved system is that windy days make for over ventilation, made worse by the extent of air infiltration into less airtight dwellings; on other days, ventilation levels may be too low. Passive stack doesn't work well in summer when the temperature difference is too small, so it is not suited to internal bathrooms – otherwise just open a window. Human perception of indoor air quality is poor, so occupant-controlled ventilation is unreliable, automated systems are taking over. In the opinion of the Passive House Institute, in a modern airtight house, natural ventilation as an alternative would require windows to be opened for five to ten minutes every three hours each day, and night, to meet the equivalent air change rate. As is the case for Passivhaus new builds, the curious anomaly is that the finished product is anything but passive, MVHR is the cornerstone of the method. MVHR systems are designed to run either at constant pressure or constant volume, whether or not rooms are occupied; although there is a growing range of ancillary devices, such as motion sensors, that could offer closer control.

For a short while, one manufacturer offered a system with demand control. Usually, the valves in each room have to be accurately set at installation, a time-consuming process because altering one throws the whole balance out. Subsequently, if extra ventilation is needed in a particular room, MVHR boosts the supply to all the rooms, whether they are occupied or empty. In this instance, though, constant fan speed was combined with humidity-sensitive extracts in each room, to regulate the air changes, reducing energy consumption, and making for both quicker and easier commissioning. But for whatever reason, this was another good idea that did not catch on. Passivhaus anticipates a lower level of heat reclaim than factory test ratings when installed. But installations that have been studied, in new, speculative houses, have found a large proportion that don't even remotely meet that standard, due to the design and installation of the ducting, the pressure loss that occurs at bends, and the losses that occur in the routing of ductwork. In retrofit properties, finding space for the equipment is difficult, and can be highly disruptive if floors are ripped up to provide a path, or if ducts are boxed in, projecting at the corners of ceilings. So, the alternatives likely to be espoused within the Future Homes Standard, either MVHR or continuous mechanical extract (MEV), are little suited to street-by-street retrofit of two-storey houses, given the impossibly difficult problem of accommodating duct runs.

Having installed a mechanical system, intended to be free from human interference, nevertheless the summertime override setting that disables the heat exchanger, will not always be able to supply the volume of air needed for cooling. So, these dwellings will need openable windows, which can of course be opened in the winter as well. It is customary in this country to open bedroom windows at night, and at least one study (MacIntosh and Steemers 2005) (Siddall 2012) has shown the energy benefits of MVHR being circumvented due to windows being opened, at night, in the winter. Much the same was observed by the research carried out in 2012 by the NHBC and published in *Today's Attitudes to Low and Zero Carbon Homes* (NHBC 2009), (NHBC 2012a), (NHBC 2012b) and (NHBC 2012c). Around 1000 occupiers were interviewed, few of those with MVHR changed the filters, and they opened windows to the same extent as those without. The other persistent problem was fan noise causing the MVHR to be turned off and windows opened. The Passivhaus process requires sufficient attenuation to overcome this problem, and measured noise levels to be acceptable at intermediate fan speeds (though perhaps not at the summertime higher speed). MVHR far from being a foolproof technology is subject to any number of constraints and conditions. Manufacturer's figures and those used for SAP calculations are theoretical and dependent on the system being properly balanced, that installation hasn't been influenced by ad hoc decisions to get around problems, and that an independent person has carried out the commissioning.

As with all aspects of retrofit, the outcomes can be beneficial or negligible, despite the capital outlay, according to the individual lifestyle of the occupants. It is familiar and usual to have air delivered from windows rather than through a vent in the ceiling. Lack of control, and unbalanced systems, lead to complaints about draughts and vents being blocked, though that is equally true of trickle vents within windows. Passivhaus has many ardent supporters due to its sophisticated software and the standards that need to be met for certification; many completed projects so far have been for adherents. But MVHR, although popular with manufacturers of fans and ductwork, does need well-informed users, not least to change the filters every few months. In city locations, the filters make for good indoor air quality, perhaps less of an advantage in future with the introduction of electric cars, although the particulates from car tyres will still be an issue since they are ultra-fine. In Finland, the country with the greatest extent of domestic mechanical ventilation in Europe, where MVHR in new developments is usual practice, there are graduated rules governing the size of filters, which are more stringent when in urban locations, even compared with Passivhaus requirements. Concern about the indoor health threat presented by dirty, smelly fans and ductwork, needs every part of Finnish systems to have access for cleaning, an aspect only now beginning to be addressed in the 2021 revision of UK Building Regulations Part F. Although ductwork may be clean when it arrives on site and is fixed, that won't remain the case if the fans are run before the house is totally clean, which given the prevalence of plaster dust may be an impossible dream. Unfortunately, maintenance has in the United Kingdom long been the victim of our 'fit and forget' culture.

The expectation is that before long all new houses constructed here will be air-sealed to the extent that MVHR will become the norm, and according to the extent of retrofit, better ventilation provision is progressively being required when dwellings are renewed. But in the same way that Passivhaus expects uniform temperature throughout the home, irrespective of whether or not rooms are occupied, MVHR delivers air to a uniform specification with minimal expectation of user involvement or infringement. The many problems that have been encountered with installations so far have resulted in every aspect of MVHR having the most cautionary label within PAS 2035. Prior to the Green Deal, separate studies by the Technology Strategy Board,

NHBC and the Good Homes Alliance (TSB 2012), (NHBC 2009a), (Good Homes Alliance 2011) and (Gap between Aspirations and Reality 2008) found widespread instances of poor MVHR design and installation, little evidence to support claims of performance, and a lack of consistency in test procedures. Complexity, which is that much greater when retrofitting, is the root cause of many problems. MVHR receives generous treatment within SAP based on assumptions about the efficiency of fans and heat exchangers, but failure to get all the components right, and rigorous commissioning, can have profound consequences if occupants turn the system off when they find that it isn't actually saving money. Brenda Boardman's hope and expectation was that all the country's housing would eventually be brought up to Passivhaus standard, or if absolutely necessary, demolished and built better. The likelihood is though that rather than the rigour of Passivhaus certification, many poorly installed, cheap MVHR systems will be found noisy, troublesome, eventually with broken fans, noxious ductwork and increasingly unnecessary as cities become greener and cleaner. The cost of certification is already proving too much for council budgets, so schemes are being completed 'to passivhaus standard,' but perhaps proving substandard. It is anyway difficult to see how MVHR will make a significant contribution to a street-by-street national retrofit programme, which needs a wholly different approach.

One might think, as with the drive to naturally ventilate other building types that the ideal system would be one with regulated natural ventilation (bearing in mind the variation in the climate drivers of wind and temperature difference) incorporating heat reclaim and of guaranteed operability for airtight buildings. SAP extension Q was established to allow the trial and adoption of novel methods of ventilation. It has long been maintained that 'demand control,' for example, the use of humidity controls on the means of extract, whether fan or passive stack, can reduce the ventilation rate, or increase it when kitchens and bathrooms are in use and generating moisture. In the case of passive stack with air supply through trickle vents, cross ventilation will tend to circumvent the route between window and stack, but if combined with continuous mechanical extract with demand control, it can overcome that problem. Alternatively, now that passive stack is a thing of the past, one manufacturer has an upgraded 'demand control' system for the retrofit of passive stacks, humidity sensitive vents are installed not only in the wet rooms but also at windows in the living and bedrooms (Aereco n.d.) (EST. 2006).

Of course, alternative technologies have been proposed. Ventive (Ventive n.d.) was developed as a wind-driven extract cowl for installation at the top of existing chimney stacks, similar in concept to the larger devices used on the roof of Bedzed (Bedzed n.d.). The Lunos (Lunos n.d.) system has mechanical vents within windows that 'talk' to one another and open and close according to the wind strength and direction. The 'supply air' windows made in France by Ridoret of La Rochelle (Ridoret n.d.) rely on air being drawn into the room between the panes. Rather than using sealed units, the air path between the layers of glass catches heat escaping from the room and delivers it back pre-heated. This not only counteracts the sensation of draughts but can achieve startlingly low U-values. Supply air windows have an unfortunate habit of fogging up on the leeward side of the building, if moist internal air escapes, rather than enters, through the windows. But combined with a strong enough extract fan, so the air path is continuous from in to out, the system provides heat reclaim. The problem with natural ventilation is that air movements can be slow; therefore, large volumes are required, so 'supply air' windows have difficulty complying with the 'effective area' requirements of the building regulations. 'Dynamic Insulation' is a similar idea, air is preheated as it is drawn in through the insulation, but it is similarly subject to the vagaries of the weather.

This debate has, however, been closed off by the 2021 revision of Part F. On the grounds that dwellings are getting smaller, and being constructed to be more airtight, the ventilation requirements have been increased. The only completely natural ventilation alternative, passive stack, has been discarded as a system though mysteriously it may still achieve compliance. To route the stack through loft space is a lot easier than threading ducts between rooms. True, it needs work on the roof but when other external works require scaffolding, such as the installation of external insulation, it is readily achieved.

Certainly, any natural ventilation system is subject to variations in flow rate according to the temperature difference inside to outside or the speed of the wind. 'Demand control' ventilation has long entailed humidity sensors at the supply side, within the window vents in habitable rooms, and humidity-controlled extract grilles to the base of the passive stacks, which lessen the likelihood of over-ventilation. Humidity controls function without electricity, they are both passive and reliable (Aereco a n.d.). This points to the urgent need for the development of domestic Building Automation and Control Systems (BACS) to boost flow rates according to demand and prevent unnecessary ventilation when it is not required. The message promoted by the building controls industry 'meter, monitor and manage' may be the route to countering the argument in favour of constantly running mechanical systems, that adequate operation of natural ventilation requires an unreasonable degree of occupant engagement. Despite 'assisted' passive stack, with intermittently operated fans within the height of the stacks to boost airflow when required, previously being accorded 'best practice' standard by the Energy Saving Trust, passive stack components are scarcely manufactured any more. Speculative house builders consider passive stack expensive, which is difficult to understand given the cost of MVHR, but of course, they are not concerned with the ongoing maintenance costs of mechanical systems.

Instead, now the only alternative for natural ventilation (in less airtight properties) is for air to be supplied by trickle vents and extracted by intermittent fans in 'wet' rooms, which won't supply continuous background ventilation, though more expensive units can reclaim heat from the exhaust. Rather than the development of passive stack 'hybrid' technology, with the inclusion of additional sensors and controls, instead within Part F 2021, the design, commissioning and testing of mechanical systems are being subjected to an ever-more rigorous regime in the hope of achieving compliance. This hopeful approach flies in the face of history. The problem is that unlike a broken boiler that will soon be noticed as temperatures drop, the human perception of indoor air quality is different from that as measured, some of the most noxious pollutants are not detectable by smell, and different individuals have varying notions of air quality.

Retrofit is so dependent on the form of dwellings, their location and occupancy that very particular solutions are necessary rather than blanket prescriptions. For example, Prasad and Patel's low-energy drying space, an alternative to electricity-hungry tumble driers, exploits the section of the building and its orientation to added spatial benefit (Prasad 2012a). The scheme came about once the architects had discovered that laundry was responsible for a high proportion of the residents' energy bills, and the possibility of converting a gloomy stairwell into a drying sunspace.

The environmental envelope doesn't have to follow the waterproof envelope. For example, on the north side of the house, normally the kitchen location in passive solar houses, an unheated larder could be made outside of the insulated rooms of the house but inboard of the line of waterproofing. Given our warming climate, revisiting Edwardian schemes for summertime open-air bathrooms is of interest, and outdoor kitchens now that much UK cooking is done on the garden barbeque (if only low energy, non-polluting barbeques existed). To exploit these potentials is to make the house, as advocated by Lisa Heschong in *Thermal Delight in Architecture* (Heschong 1979), a sequence of thermal environments where the enjoyment of insulated warmth is perceived by reference to ambient conditions, i.e., living with nature. This is a different sensibility to that of mechanical control, where the whole dwelling is designed as a single heated volume, of uniform thermal conditions.

In many ways, ventilation is the Achilles heel both of EnerPHit, and of the pressing need for large-scale low-energy retrofit. MVHR systems can only be installed all at once, with huge internal disruption, so they aren't conducive to step-by-step retrofit, which needs to align with maintenance schedules and the lifetimes of existing components, the only way of making costs affordable. There is room here for architectural invention and ingenuity, new and better methods are urgently needed.

References

Aereco. n.d. Available online: https://www.aereco.co.uk/ventilation/ventilation-systems-uk/natural-ventilation/ (accessed 5 October 2022).

Aereco a. n.d. Available online: https://www.aereco.co.uk/ventilation/humidity-sensitive-ventilation/ (accessed 22 May 2022).

Bedzed. n.d. 'Beddington Zero Energy Development'. Available online: https://www.zedfactory.com/bedzed (accessed 6 July 2022).

Boardman, B. 2007. *Home Truths*. Oxford: Environmental Change Institute, University of Oxford.

Boardman, B. 2012. *Achieving Zero – Delivering Future Friendly Buildings*. Oxford: Environmental Change Institute, University of Oxford.

Brand, S. 1994. *How Buildings Learn: What Happens after They're Built*. New York, NY: Viking.

EST. 2006. *GPG268 Energy Efficient Ventilation in Dwellings – a Guide for Specifiers*. London: Energy Saving Trust. (PSV inline fans previously enabled passive stack to achieve 'best practice').

'Gap Between Aspirations and Reality' (2008). *Architects Journal*: 6 March.

Good Homes Alliance. 2011. *Ventilation and Good Indoor Air Quality in Low Energy Homes*. London: Good Homes Alliance.

GOV.UK. 2006. 'English House Condition Survey 2006'. Available online: https://www.data.gov.uk/dataset/f2f2fd99-cea6-41fb-ade0-1b4f22bb709b/english-house-condition-survey (accessed 2 October 2022).

GOV.UK. 2008. 'Heat and Energy Saving Strategy'. Available online: https://assets.publishing.service.gov.uk/government/uploads/system/uploads/attachment_data/file/243625/9780108508158.pdf.

Heschong, L. 1979. *Thermal Delight in Architecture*. Cambridge, MA: MIT Press.

Inventer. n.d. Available online: https://www.inventer.eu/ (accessed 10 July 2022).

Lerpiniere, A. 2022. 'Clean Air Action', *RIBA Journal*, May: 80.

Lunos. n.d. Available online: https://www.lunos.de/en (accessed 22 May 2022).

MacIntosh, A., and K. Steemers. 2005. '*Ventilation Strategies for Urban Housing: Lessons from a POE Case Study*'. Building Research and Information (January – February 2005) 33(1): 17–31.

NHBC. 2009. *A Practical Guide to Building Airtight Dwellings*. London: NHBC Foundation.

NHBC. 2009a. *Indoor Air Quality in Highly Energy Efficient Homes – A Review*. London: NHBC Foundation.

NHBC. 2012a. *Today's Attitudes to Low and Zero Carbon Homes*. London: NHBC Foundation.

NHBC. 2012b. *A Survey of Low and Zero Carbon Technologies in New Housing*. London: NHBC Foundation.

NHBC. 2012c. *Mechanical Ventilation with Heat Recovery in New Homes*. Ventilation and Indoor Air Quality Task Group, London: NHBC Foundation.

Prasad, S. 2012a. "Penoyre and Prasad's Retrofit for the Future Project." *RIBA Journal* 1 (November): 70.

Ridoret. n.d. Available online: https://www.fenetre-enr.fr/pdf/2021-Brochure-EnR.pdf (accessed 22 May 2022).

Siddall, M. 2012. 'Passive Ventilation Climatic Consideration', *Green Building (32), Summer edition*.

TSB. 2012. *Retrofit Strategies: Key Findings – Retrofit Project Team Perspectives*. London: Technology Strategy Board.

Ventive. n.d. Available online: Ventive: https://ventive.co.uk/ (accessed 22 May 2022).

10 Anticipating the Green Deal

Following the 2008 Climate Change Act, passed in response to the target set at Kyoto, and the strategy documents that followed, retrofit anticipation was at its highest. The 2009 Low Carbon Transition Plan stressed the importance of a 'whole house approach' to retrofit, and the need for houses to receive a comprehensive series of improvements, able to be paid back through energy bill savings over the lifetime of the building. The intention being that the needs of the dwelling should be assessed as a whole, happening in the correct order so as to avoid disruption; such was the basis of the Green Deal. The following year, the government strategy 'Warm Homes, Greener Homes' anticipated that following the completion of simplest measures, loft and cavity wall insulation, seven million eco-upgrades would be completed by 2020; entailing solutions tailored to individual properties (given that the energy performance of every household is different), dealing with the whole house, and incorporating renewables and low-carbon energy supply. This was highly ambitious given the age of the stock, and that two-thirds of homes built before 1919 were owner-occupied. At that time, 60 per cent of energy consumption was for space heating and only 3 per cent for both cooking and consumer electronics.

To find ways forward, the Retrofit for the Future Competition from 2009, aimed for an ambitious 80 per cent carbon reduction (that few of the contestants achieved). But they came up with a variety of technical approaches, which, although expensive, suggested the possibility of economies once replicated at scale. The results of the competition encouraged belief that reductions of between 50 per cent and 70 per cent could generally be achieved within retrofits. Given the wide variety of house types involved, each with their own initial energy benchmark that was unknown, and had to be calculated, the 80 per cent target was fairly meaningless, and was much easier to reach with initially poorly performing properties. The recorded problems remain prevalent, including the difficulty of obtaining appropriate products, most of which are imported, obtaining monitoring equipment with consistent performance and working with residents who disliked the disruption – often requiring them to be decanted into temporary accommodation.

The results of the competition were summarised by the Institute of Sustainability (IfS 2011). Each entry to the competition had a budget of £150,000, far in

DOI: 10.4324/9781003564997-13

excess of everyday budgets, the focus was on innovation. Indeed, it enabled ambitious approaches, for example, one of the most publicised projects, in Princedale Road, West London (Davis 2012) (Baeli 2013), introduced a steel frame behind the existing solid front wall, essentially a new building inside the old. The harsh reality was that at that time the average household energy bill was £1000 per annum (£600 gas and £400 electricity), so even if a very optimistic 75 per cent reduction could be achieved, the maximum saving per year would be £750. The Green Deal funding regime, the so-called Golden Rule, required that energy-saving measures should pay for themselves over the course of their lifetime; in those terms, the amount spent on Retrofit for the Future homes was many times in excess of feasible expenditure.

The Institute's response was to seek savings. On the one hand, energy efficiency adds comfort, healthier conditions, amenity and thereby value to a property, which particularly benefits owner-occupiers. Larger-scale retrofits could achieve economies of scale, by developing supply chains and working practices, but also by reducing the scope. For example, increasing the thickness of insulation to meet higher levels of carbon reduction entails ever-increasing pay-back times. If 50 mm of insulation is used rather than 250 mm, the payback is reduced by the square of the thickness, i.e., 25 times less. Cheaper systems were envisaged, such as continuous mechanical extract rather than mechanical ventilation heat reclaim (MVHR). Whilst the problem of affordability remained, subsequent rocketing gas prices, and the costs to consumers of decarbonising the grid, effectively make payback times now that much shorter. But one of the outcomes of the competition was that occupants were now more comfortable and enjoying their higher room temperatures, thereby increasing payback times yet again. The complexity of the task ahead was only too evident by time Retrofit for the Future competition came to an end.

The report's conclusions airily skated over this central problem. Although the Green Deal was clearly not going to pay for whole-house retrofit, it was suggested that perhaps the easiest house types should be tackled first and that another pilot project should be launched by the government to build UK-based supply chains and reduce costs. And indeed, there was eventually a second competition to try and break the log jam (GOV.UK. 2019).

On the runup to the Green Deal in 2009, Gentoo was one of five social housing providers to be given funding for a prototype Pay-As-You-Save (PAYS) pilot scheme. In 2010, Gentoo carried out an extensive tenant engagement programme (Gentoo 2009) involving 600 of its homes to determine their willingness to pay for energy-saving measures. Buoyed up by experience thus far, the following year saw the launch of its own larger-scale Green Deal Pilot scheme, the 'Energy Saving Bundle' for the retrofit of 1200 homes in Sunderland. This led to Gentoo becoming the first social housing landlord to join the Green Deal Finance Company, and subsequently an accredited Green Deal Provider. So far, so good. But the Energy Saving Bundles were subsidised by funding from CESP. Individual dwellings were assessed for likely savings, and a bundle of measures was included where appropriate, including double glazing and solar PV, boiler upgrades were provided free. The anticipated savings were reduced for some 'rebound' increase in comfort, and charges were set just below the level that, if exceeded, would break the 'Golden Rule.' On completion, customer questionnaires were evaluated, and energy meter readings were analysed, data loggers were installed in some of the properties that were also pressure tested. Word-of-mouth recommendations between residents followed on from the engagement programme, the retrofits were warmly received, and the payments, which could be as an addition to rents, or directly to the gas company, seemed satisfactory. Gentoo emphasised the importance of a community approach. One year on from retrofit, it was not clear if their customers were showing any benefit on their energy bills.

Also in 2009, Arup published their study into the feasibility of retrofitting privately-owned homes in Bristol (Arup 2009), which constitutes 93 per cent of housing in Bristol City. The stock was divided between types of construction, to which different packages of improvements were modelled in SAP from simplest measures such as energy-saving lightbulbs and draughtproofing, to the more ambitious addition of panels on roofs and external insulation. For consistency of pricing for homeowners, and maximum financial benefit, it was advocated that whole packages should be delivered at the same time to cut down on overhead costs; multiple homes in the same neighbourhood being retrofitted concurrently was the way to achieve significant economies of scale. The report was offered as support to the then current government consultation on the Heat and Energy Saving Strategy, which aimed for all the country's stock to have a 'whole house' retrofit by 2030.

Preparatory to the Green Deal, the mechanism that was supposedly going to lead to countrywide retrofit, a range of uncoordinated studies were carried out by array of organisations, such as the Institute for Sustainability and the Zero Carbon Hub, both of which are now defunct.

Rickaby Thompson (Derby 2013) carried out studies into a number of dwelling types using simulations to assess the extent of basic insulation measures necessary to achieve SAP80, the top of EPC band C, that was a proxy for the 'affordable warmth standard.' They went on to evaluate extra measures to reach a carbon reduction of 60 per cent, external door and window replacement, or programmable thermostats and PVs; the cost of the two packages, at least for older types, was about the same.

ESRU at the University of Strathclyde (ESRU 2005) completed a similar investigation in Scotland, classifying the country's housing stock into groups, and deriving pay-back times for alternatives. Most of the varieties of insulation, loft and wall, and upgrading heating systems, paid for themselves in less than five years; double glazing was the longest at seven years (if the windows were due for replacement anyway). Their model, if applied to the whole housing stock, would result in a 33 per cent reduction in total national annual heating demand, at completion of just basic levels of insulation and draught-proofing, in the three most common types. Their model showed the percentage increasing to 50 per cent at completion of a higher standard of the same measures, if applied to every home in Scotland.

Parity Projects (Parity Projects n.d.) has, since 2005, been developing a database and analytical methods applied to retrofit proposals, based on the reality that each home is different. A number of consultancies, run for profit, have increasing knowledge that is commercially sensitive. The plethora of acronyms, methodologies and standards is the inevitable outcome of market competition. Parity has its own evaluation system (Home Energy Master Plan, HEMP) distinct from the Home Energy Rating (NHER) that can be used to estimate fuel use, costs and carbon emissions. In addition, HEMP is linked to their in-house databases for detailed simulation modelling, it can be calibrated against bills and used as the means to advise on alternative retrofit strategies. CarbonBuzz, the RIBA/CIBSE platform, uses a different method for calculating primary energy use and CO_2 emissions, so it gives a different result to Parity Projects, but in both cases radically different from the in-use measured data. Various construction companies have their own databases used to advise prospective clients; is this really the way to address the beginnings of a street-by-street UK-wide retrofit programme?

There was some initial success, Green Deal assessments were underway and there were hopeful signs of public interest. The Department of Energy and Climate Change (DECC) hoped for a saving of 4.5 million tonnes of CO_2 per year, which would also assist 100,000 low-income households, though the parliamentary select committee could find little evidence of how that was to be achieved.

The National Refurbishment Centre 2012 report *Refurbishing the Nation – Gathering Evidence* (National Refurbishment Centre 2012) concluded that retrofit was beset by complications and costs that exceed initial estimates. An Arup presentation, made at Ecobuild 2013, preparatory to the Green Deal, outlined the ambitious goal underlying the Climate Change Act, a 34 per cent reduction in CO_2 by 2020. Some of the issues were highlighted: The country's wide range of dwelling types with their own individual retrofit requirements, the rebound effect whereby occupant behaviour could undermine reductions in consumption, yet the massive reduction in unit costs that might be achieved through retrofit at scale. All of these questions and opportunities have to be understood as a body of knowledge, and a database of solutions, before street-by-street retrofit commences. The problem was, of course, the necessity to spend in the short-term for long-term gains, the conundrum that has beset all efforts to get a national programme underway.

Problem #5 Airtightness

As insulation standards have increased, infiltration has assumed greater importance as it is now responsible for an increasing proportion of total heat loss, thus the current motto 'build tight, ventilate right.' Airtight dwellings could be extremely hazardous to human health if they aren't equipped with adequate ventilation. Even so, toxic off-gassing from materials is a hazard locked into new construction, requiring additional fresh air over and above that needed for respiration. This is a particular concern in the United States, where indoor air quality (IAQ) is listed as the fifth greatest risk to human health. Many pollutants can be three to five times greater in concentration indoors than outdoors, though volatile organic compounds (VOCs) and formaldehyde are possibly 20 times as concentrated, making IAQ as great a concern as high levels of city pollution. This is a topic of sufficient concern that Google has reacted with formation of a healthy building tool, an approved materials database to avoid future problems in their own building stock. Builders in the United States have been nervous about tightly sealed construction, so ASHRAE in 2014 published Standard 62.2 (ASHRAE 2014) that specifies minimum ventilation rates.

Emissions of organic compounds from building materials have been an issue for many years resulting in product labelling requirements set by the EU. In the United Kingdom, levels of permissible CO and NO_2 are consistent with WHO stipulations for long-term, but not short-term exposure. The WHO also defines safe concentrations of formaldehyde, tetrachoroethylene, naphthalene and radon, which are not currently included in Part F performance standards. For some compounds, such as benzene, the WHO provides no safe standard but gives a health risk factor according to their concentration. UK regulations for outdoor air quality include levels for particulates, sulphur dioxide, benzene, 1,3-butadiene and polyaromatic hydrocarbons (PAHs), also nitrogen dioxide and carbon monoxide, which are as yet only under consideration for indoor purposes. Not surprisingly, given the uncertainty around these issues, regulations regarding airtightness have not been tightened with the same rigour as, for instance, those concerning insulation, although under 2021 regulations all new houses have to be pressure tested, not just a sample as previously.

In the United Kingdom, the government in 2019 published a survey of ventilation levels in new homes (GOV.UK. 2019a), 80 homes were studied across seven sites in England, many of them didn't meet the building regulations 2010 Part F requirements for fan performance or provision of trickle vents. The sample houses were chosen from a variety of locations, and the buildings were extensively monitored. The results were startling. Only two types of systems were analysed: manually operated fans in wet rooms and what was found to be a favourite type of installation amongst housebuilders, comprising a third of the total, decentralised constant mechanical extract (dCME), air being drawn in through trickle vents and extracted via constantly running fans. MVHR was not part of the survey, although that has also become widespread in recent times. Very few of the dwellings met the regulations, often for familiar reasons, trickle vents blocked because of draughts or obstructed by bedroom curtains. The vents were often left open in bathrooms, which short-circuited the extract rather than drawing air from the rest of the house. But the fans themselves were a problem, often with insufficient rate of extract, wrongly installed, or, in the case of dCME, used intermittently by householders who used the isolator as an on/off switch.

Although the monitors found evidence of poor air quality, this wasn't confirmed by the users, demonstrating the extent to which human perception of indoor pollution is untrustworthy. Not surprisingly, there was plentiful incidence of condensation and mould. Optimal relative humidity is between 30 and 60 per cent; a metric of significance because too low or high a level may also be associated with bacteria, viruses, allergic rhinitis and asthma, whereas dry conditions suit fungi and dust mites. Not only is England frequently wet, but the widespread practice of drying clothes indoors in winter exacerbates potential problems.

> The root of all evil is water… Rain is only the most obvious source of the problem. More pernicious now in most buildings is internally generated water vapour. Every year since the energy crisis of 1973, buildings have been made more airtight and better-insulated to save on energy costs. But keeping in the nice warm air (or the nice cool air in hot seasons) meant also keeping in all the moisture that humans, kitchens, and bathrooms constantly exhale.
>
> (Brand 1994, 114)

In relation to airtight buildings, moisture is indeed the root of all evil. In 2019, Glasgow's Mackintosh Environmental Architecture Research Unit (MEARU) monitored 30 dwellings (MEARU 2019) (MEARU 2016), a mix of old and new, and found that typically a load of washing releases about two litres of water. When clothes were left to dry in ill-ventilated rooms, which was found to be the norm, mould and dust mites flourished, underlying causes of asthma and eczema. A ventilated tumble dryer has to be ducted to the outdoors and uses expensive energy, as does opening a window during the winter. There has been an increasing incidence of high humidity and poor air quality in airtight homes, due to inadequately executed details for insulation, vapour barriers and due to MVHR being badly installed. MEARU's survey of the 'Glasgow House' demonstration project similarly found poor airflow rates; air was being extracted from the kitchen and bathroom and supplied to the hall, but unable to find its way into other rooms due to the intumescent fire-stripped doors. The 2021 revision of Part F has sought to overcome some of these problems, restricting the use of flexible ducting in MVHR systems, requiring doors to be undercut to facilitate air movement, and introducing regimes for systems to be commissioned and tested in use. Many of the problems, the lack of mesh between different building trades, insulation and windows, airtightness and ventilation are due to a lack of joined-up thinking. The role of the Retrofit Coordinator is intended to address this ongoing difficulty.

An airtightness layer has to be impervious to wind-driven rain and air, whilst allowing movement of water vapour. Sheet materials joined with adhesive tape can be applied internally or externally as can airtightness paint. Internal application has the advantage of being applied independently of weather conditions, perhaps following the line of the vapour control layer, but with the disadvantage of being incomplete since partitions and floors abutting the outer wall breach the air enclosure. Applied externally as sarking, many of these problems are overcome and the occupants are subject to that much less disruption, but the sarking is fixed outdoors in variable weather conditions. Putting the airtightness layer to the outside of the structural fabric, requires the work being done out in the open air, making an already workmanship-dependent technology doubly difficult. Although these clever materials have been subject to accelerated tests that suggest a long lifespan, they are nonetheless highly workmanship dependent, and whilst being flexible, whether they can withstand long-term movements, particularly if the substrate is timber, remains to be seen.

Because the long-term durability of tapes and seals is unlikely, the Institute of Sustainability recommended that seals around windows, doors and skirting boards should be periodically checked and, where necessary, replaced with a concluding pressure test (IfS 2011a). But what of all the hidden tapes, sealants and airtightness tape? Anyway, within the UK's 'fit and forget' tradition, will any of this come to pass? Leakage points can be detected using smoke pencils when the building is being pressure tested, but this is not likely to be helpful if the airtightness layer is already buried within the construction, and even less so if the layer is towards the outside. On the other hand, in existing buildings, the cause of leakage, such as service entries, can be hidden within or behind parts of the construction, making the task impossible. Given that most retrofit situations involve masonry-built dwellings, with wet plaster walls, a separate airtightness layer may be an unnecessary expense once the obvious leakage points around windows and service entries have been sealed. In addition, foil-faced insulation board, with aluminium foil-taped joints, will be airtight as well as acting as a vapour barrier for IWI. Whilst the most common EWI, using external render over insulation boards, is similarly airtight. So, in many retrofit situations, the design of the airtightness layer will be an issue already dealt with; solutions have to be individually tailored to suit different homes. Nonetheless, given the prevalence of poor workmanship that leaves airgaps, and Building Control that has become less than rigorous, can we assume that work that ought to last decades actually will? This is a subject, the longevity of airtightness, that has been little-studied, but one recent paper suggests that a 20% reduction can be expected within two years after completion, after which the level begins to stabilise (Leprince 2022).

Project Calebre (University of Loughborough 2013), a consortium of universities, investigated a number of technologies for retrofit prior to Retrofit for the Future. A test house, replicating a 1930s semi-detached, was used to progressively increase the airtightness of the building from its original 15.6 m^3/m^2.hr (at 50 Pa) down to a final maximum achievable value of 4.8 m^3/m^2.hr that included lifting the ground floor coverings to install an airtightness membrane. The project's final report in 2013 pointed to the difficulties experienced in achieving the result, and the care and attention to detail required by the workforce. The team then went on to investigate the use of MVHR: 'since the effectiveness of an MVHR system depends on the correct balance between heat recovery efficiency, fan efficiency, air flow rate and building airtightness, there is a technical challenge in using MVHR for retrofit.' For a 'minimum' efficiency system with a heat recovery of 70 per cent the airtightness would need to be 3 m^3/m^2.hr to save energy relative to natural ventilation, and 1 m^3/m^2.hr to save carbon because the fans run on carbon-intensive electricity. If a more efficient, at that time best practice, version with 85 per cent reclaim was used, the figures were 5 m^3/m^2.hr and 3 m^3/m^2.hr, respectively, but at that time the results challenged the viability of MVHR in retrofit. The report recommended that only the best of products should be used. The results were only a snapshot, tests in other house types were recommended and now, some years later, higher efficiencies are being achieved by manufacturers' factory tests, but the payback periods may be very long.

The belief that all problems are amenable to scientific analysis, and engineering solutions, is the same philosophy that has held sway since the Enlightenment and underlay Corbusier's 'Machine for Living in.' The Building Regulations similarly aim for a degree of certainty, which is rarely achieved in practice. True enough, a certified method of construction such as Passivhaus that aims to avoid all the usual mistakes within construction and arrive at a result closer to the design intention, must be applauded. However, one of the Retrofit for the Future entries, Paul Davis and Partners Princedale Road scheme (Davis 2012) (Baeli 2013) illustrates what is required to meet the EnerPHit target of 0.6 m^3/m^3.hr at 50 Pa ($\leq$ 0.6 ACH @ 50 Pa) a standard way in excess of the leakage requirement in Building Regulations L1A. The roof and all the external walls, and the lower floor, were clad with OSB board sealed to the windows and doors with airtight tape, a degree of effort that can hardly be expected of a rapid street-by-street retrofit programme, least of all one with occupants staying in their homes.

But on the other hand, there are critical voices that would prefer homes that are enjoyed as environments, engaging with the changing weather and seasons. In England, this has traditionally involved a lifestyle where pullovers are put on and taken off. Not only variable temperatures impinge but also different rates of airflow, some days may be breezy with noticeable draughts, other days less so. At a time when the future climate is so uncertain, flexibility is necessary to accommodate a shortening heating season of warm and wet winters. But the drive, within new build, is to take away all response to the environment of rooms and put it into the hands of mechanical devices. Perhaps the hope for the future is with control systems that vary temperatures, lighting and ventilation according to individual preferences on one's mobile phone. If, as a result, rooms that are unoccupied can be heated less, the savings may be a lot more significant than heating super-insulated houses to the same temperature throughout. Similarly, more sophisticated methods are needed for the design of naturally ventilated systems that accept a level of over-ventilation on some days and rather less on others, combined with dynamic analytical tools that anticipate air-quality changes over time. Airtight buildings are a potential cause of huge problems, unless ways of ventilation can be developed that aren't as evident, and annoying, as fans and trickle vents, just part of the background of life as was infiltration into the air-leaky buildings of the past. MVHR doesn't fit that description even though it may be the best we have so far. IAQ has the makings of being the next regulatory scandal, particularly given the current context of increasing fuel poverty, and the cost of running mechanical equipment in increasingly airtight homes.

> Whereas competent sealed buildings lull us with their "perfect" climate, and incompetent ones drive us crazy with their uncontrollable heats and colds, a draughty old building reminds us what the weather is up to outside and invites us to do something about it – put on a sweater; open a window. Rain is loud on the roof. You smell and feel the seasons … Such buildings leave fond memories of improvisation and sensuous delight.
>
> (Brand 1994, 33)

References

Arup. 2009. *Forum for the Future: Bristol Retrofit, Final Report*. London: Ove Arup and Partners.
ASHRAE. 2014. *Standard 62.2*. Available online: https://www.ashrae.org/search?q=Standard%2062.2 (accessed 10 July 2022).
Baeli, M. 2013. *Residential Retrofit 20 Case Studies*. London: RIBA Publishing.
Brand, S. 1994. *How Buildings Learn: What Happens after They're Built*. New York, NY: Viking.
Davis, P. 2012. '100 Princedale Road for Octavia Housing, Prize winner' Retrofit Awards, *Architects Journal*, 1 June.
Derby, K. 2013. 'Targets for Retrofit: Achieving SAP0 and C60'. Kathryn Derby, Rickaby Thomson.
ESRU. 2005. 'Thermal Improvement of Existing Dwellings'. ESRU University of Strathclyde. Final Report to the Scottish Buildings Standards Agency, January.
Gentoo. 2009. *Retrofit Reality Dissemination Report*. Sunderland UK: Gentoo Group.
GOV.UK. 2019. *Whole House Retrofit Innovation Competition*. Available online: https://assets.publishing.service.gov.uk/government/uploads/system/uploads/attachment_data/file/890020/Whole_House_Retrofit_Supplier_Event_Webinar_June_2019__withdrawn_.pdf.

GOV.UK. 2019a. *Ventilation and Indoor Air Quality in New Homes.* Available online: https://www.gov.uk/government/publications/ventilation-and-indoor-air-quality-in-new-homes (accessed 10 July 2022).

IfS. 2011. *Low Carbon Domestic Retrofit: Retrofit Insights – Perspectives for an Emerging Industry – Key Findings: Analysis of a Selection of Retrofit for the Future Properties.* London: Institute for Sustainability.

IfS. 2011a. *Low Carbon Retrofit Guides – Chapter 5: Managing Low Carbon Retrofit Projects and Chapter 6: Improving the Building Fabric.* London: Institute for Sustainability.

Leprince, V. 2022. *Durability of Building Airtightness*, AIVC Technical note TN71. Available online: https://www.aivc.org/resource/tn-71-durability-building-airtightness (accessed 12 June 2023).

MEARU. 2016. *MacKintosh Environmental Architecture Research Unit: Post Occupancy Evaluation of GHA's Demonstration House 'The Glasgow House'.* Available online: https://mearunet.files.wordpress.com/2016/12/glasgow-house1.pdf.

MEARU. 2019. *MacKintosh Environmental Architecture Research Unit: The Environmental Impact of Domestic Laundry.* Available online: http://radar.gsa.ac.uk/6852/ (accessed 10 September 2022).

National Refurbishment Centre. 2012. *Refurbishing the Nation – Gathering the Evidence.* Available online: https://www.scribd.com/document/145651193/Refurbishing-the-Nation-Gathering-the-evidence/ (accessed 6 June 2022).

Parity Projects. n.d. Available online: https://parityprojects.com/ (accessed 23 May 2022).

University of Loughborough. 2013. 'CALEBRE – Consumer Appealing Low Energy Technologies for Building Retrofitting', September. Available online: https://www.cibse.org/media/cw4nmpyi/calebre-cibse-presn-v3-jul13-d-loveday.pdf.

11 Retrofit Comes to a Halt

The Green Deal was thought to be the mechanism for all this to happen. The idea of it was simple, nobody pays up front, houses are insulated etc. and the cost is set against the house like a mortgage, but monthly gas bills stay the same, so the reduction in energy use progressively pays off the debt. The so-called 'golden rule' was that the energy reductions must be sufficient to pay off the debt within a reasonable length of time. Some of the projects preparatory to the Green Deal, such as Birmingham Energy Savers (BES 2012) used the same formula, the reduced energy bills being set against the costs of retrofit with high-interest loans being paid off over time. But as was the case for the Green Deal, this debt was set against the property effectively reducing its value at sale, which is very unappealing for property owners. The Birmingham project was wound up in 2015, with only 16 homeowners having taken up the opportunity. Different approaches are needed for social housing and privately rental properties. For example, where 'warm rents' have been applied by housing associations, a surcharge for living in an energy-efficient home, the housing provider has to pick up the bill when dealing with the problem of rent defaults.

Of course, some houses are more difficult and expensive to improve than others, which is where the financing started to get complicated. Getting the process started requires 'sticks and carrots,' the stick is the law, tougher building regulations requiring retrofit works to be carried out, particularly whilst other work is being done on the property. More recently, it seemed that in relation to the rental market improvements were finally underway. Poor accommodation, anything below an Energy Performance Certificate 'C' rating, would not be able to be let after 2028. But then in 2024, a government U-turn scrapped this previous good intention.

Before long the Green Deal had collapsed, before it had really got off the ground. This was warmly anticipated by the architectural press, who were appalled that a Green Deal assessor, qualified after just a three-day course, could advise homeowners on how to retrofit their properties using a range of technologies from insulation to photovoltaics, double glazing to biomass (Murray 2013). The fear was that a rash of uPVC windows and insulated render would be environmentally detrimental. Homes would be unsaleable, given the high interest rate and loans that were to be attached to the properties, not the owners. Effectively,

DOI: 10.4324/9781003564997-14

the process had been handed over to vested interests whose sole concern was making a profit. Fortunately, this ill-thought scheme came to nothing, but its demise meant the end of any prospect of nationwide retrofit for another ten years.

The failure of the Green Deal, inherited from Labour by the Coalition Government, with badly thought-out financial inducements and half-hearted promotion by a disinterested administration, led as a result of concerns about 'cowboy builders' to the beginnings of PAS 2030, a 'Publicly Available Specification' for energy efficiency measures in existing buildings. ECO, the Energy Company Obligation, became the successor to CERT and CESP, funded by energy supply companies, for installation of measures to comply with the new PAS 2030. The companies' achievements within CERT and CESP had been very variable, some lived up to their initial commitments, others not.

The requirements for ECO are more complex than their predecessors; for example, the methodology for carbon savings calculations that have to be submitted to Ofgem. Either SAP (Standard Assessment Procedure) or the Reduced data Standard Assessment Procedure (RdSAP) can be used, but in conjunction with a number of approved software packages that each provide different answers. Ofgem has consequently found fault with many of the results, with completed schemes not able to be wound up for the lack of a correct carbon saving score.

Problem #6 Renewables

At least until 2025, the expectation is that gas will remain the main heating fuel, rather than electricity, while the carbon intensity of electricity remains high. The natural gas system will be decarbonized through the addition of green gas from anaerobic digestion thus prolonging the period of its carbon acceptability. Beyond 2025, the need for space heating will disappear as properties are made low-energy or brought up to Passivhaus standard.

(Boardman 2012)

Now we have reached an economic tipping point. By good fortune, whilst the climate crisis becomes ever more urgent and the rate of carbon dioxide emissions worldwide continues to grow, the price of renewable energy is now undercutting fossil fuels; for example, in the United States, solar generation is now cheaper than all other forms of electricity production. The intermittency problem of solar and wind is closer to a solution as a result of new technologies such as advanced types of batteries, and combined cycle gas turbines that can be switched on and off, to provide backup power to the grid. The oil-based economy is on its way to being in the past, no time to be lost if the remaining hydrocarbons are to be left underground. Transitioning from a fossil fuel-based centralised energy system to a new world of renewables and decentralisation, was always going to be difficult given the vast size of the existing infrastructure, and the influence of vested interests. Until the recent gas price crisis, the related goals, switching to low carbon, affordability, and security of supply were all pulling in opposite directions. But the context has changed, for example, in May 2021 the European Commission published its strategy to end the EU's dependence on Russian oil and gas. The €210 billion plan entails a vast scaling-up of renewable energy generation, including doubling the installed capacity of solar panels in the EU by 2025. The EU currently intends producing 40 per cent of its electricity from renewables by 2030, but the Commission's proposal is to increase the target to 45 per cent.

In the United Kingdom, for individual dwellings there will only be limited opportunities for small-scale hydro, biomass, CHP and lesser renewables, so in most cases the likely mix will be solar panels on the roof and heat pumps, both of which, to make sense, require prior retrofitting to bring the fabric up to scratch. For most households, or at least those with mains gas, the choice so far has been straightforward, conventional boilers have been cheaper to run and far cheaper to buy than heat pumps. Given the limitations on available space around most properties, the installation of Ground Source Heat Pumps (GSHP) would seem to be a limited option compared with air source ASHP, albeit that GSHP is more efficient, if more expensive. Perhaps hydrogen could be part of the mix, but that is so far off as to make little impact in the critical decade up to 2030. The much-vaunted future of heat pumps will depend on their efficiency, studies carried out by the Energy Saving Trust (EST 2010) demonstrated that the Coefficient of Performance advertised by manufacturers was rarely achieved due to poor design, and installation, and users' difficulty with the controls. EST field trials showed that a wide range of performance could be expected, according to the design and its installation, depending on individual circumstances such as the size of existing radiators, the anticipated demand for heat and hot water, and interactions with other technologies such as solar thermal collectors. There are longstanding issues with ASHP, which it is hoped will be overcome with their widespread introduction: their poor performance at lower temperatures requiring auxiliary and costly backup; their need for de-icing by reversing the cycle with further energy disadvantage; that they won't operate if covered in snow; not to mention that they are noisy, large and ugly and may not work with existing radiators. It makes for a complex system that may not be suitable for all UK locations, needs expertise and rigorous quality control (as certified by Passivhaus) but requiring for wholesale implementation, a skill base that doesn't yet exist.

However, a decarbonised future must be reached because the grid at present is without the ability to support the switch to electric vehicles and heat pumps, which implies two to three times current capacity. Additionally, by 2030, 95 per cent of electricity has to be low carbon, thereby overcoming the country's insecurity of fossil fuel supply. Local on-site generation into the low voltage grid is imperative as is increased energy storage. The grid has to transition to become a 'smart grid' capable of balancing central supply with local supply, and with storage matched to local demand. This localised grid will increase its capabilities by better balancing demand and supply across the day. Advantageously, local generation reduces transmission losses.

On the UK's crowded island, most flat land is farmland; solar farms are not as readily constructed as on a desert in Australia, lengthy planning enquiries and local objections have slowed the process. But the nation's supply of houses with roofs is readily available as sites for panels and being 'permitted development' are more easily utilised, except of course in the case of listed buildings and conservation areas. So, domestic PV electricity production will, together with the replacement of boilers with heat pumps, be central to de-carbonisation of the grid. Photovoltaics for electricity generation, and heat pumps for its consumption. Add into the mix the extra demand when charging-up electric cars and the equation becomes complex. Enthusiasm for small scale wind power lasted for a while, but now most of the in-town turbines have disappeared (although, in the right location, a small turbine 9 m tall would likely produce about the same power output as 20 houses with PV). PV is best suited to homes in constant occupation, the electricity generated being used on-site, thereby helping to cover day-to-day costs. Important given the recent exodus from offices, increasing numbers of people preferring to work from home.

Although excess electricity can be sold back to the grid, different rates apply between companies and different areas of the country. Back in 2020, one third of the UK's energy was being provided by renewables, but only just short of 4 per cent was due to solar, and four times as large a percentage provided by wind power. In July 2022, solar was responsible for 7 per cent of electricity generation and 19 per cent wind, totalling 26 per cent. Last winter, the proportions reversed with wind generation hitting 26 per cent and solar yielding nothing. In total, renewables, including biomass and hydro, are now responsible for 35 per cent in the winter and 33 per cent in summer.

Of course, into the mix must be added that developed technology, rooftop solar hot water (including the various types of hybrid panels), though more of an option for larger families with a greater hot water requirement. All of which highlights the need for design integration of mechanical and electrical systems when retrofitting. The controls for solar hot water are more complex than PV and can be compromised if poorly installed, and if electrics and plumbing aren't well aligned. Monitoring and controls could make a great contribution to energy efficiency, by factoring in the demand profile of individual rooms. Domestic retrofit is still a way from enjoying the intelligent controls, and smart metering, of Building Automation Controls Systems (BACS, the 'brain' of the building), and for their adequate recognition within the regulations. Indeed, there is no clear path towards even low-tech methods that could be used to identify households in need of energy advice, such as a way of readily comparing estimated energy usage with meter readings.

Complication will further increase if rainwater and grey water harvesting systems enter the mix. The requisite skills required are not yet available in the United Kingdom, unlike on the continent where these systems are common, with the ability to determine, for example, whether underground tanks can be economically installed. By comparison, waste water heat recovery (WWHR) is relatively easy to fit, but to date, it has not caught on, another technology that could become a standard component, but for lack of market push.

As electricity costs grow ever higher so does the appeal of PV panels. The prevailing wisdom is that to optimise the payback time, the number of solar panels needs to be geared to current household energy demand, the addition of more panels would not cover the initial investment, even over a two-decade period. With the switch to heat pumps and electric cars, the area of rooftop solar per house can be expected to increase a lot, in comparison with the limited number of panels that have been usual to date. The danger is that ageing installations will not be rapidly replaced with the requisite extra capacity, particularly since, unlike a gas boiler, there is no precise cut-off date where panels become obsolete. An average solar panel will lose about 10 per cent of its efficiency in the first 10 years, and 20 per cent by the time it is 25 years old. Most rooftop solar panels come with a warranty of between 25 and 30 years, but most will go on generating electricity, albeit less efficiently, some have been known to keep working for nearly 40 years. Solar panels can be recycled, and are in fact prohibited from being dumped, since all solar panels contain small amounts of toxic waste. They are nonetheless a potentially valuable resource, it is estimated that by 2050, solar panels will have a recoverable value of £11 billion.

A survey of estate agents by *Which magazine* (Ingrams 2017) suggested that the addition of PV panels had little bearing on the sale value of the property. Building Integrated PV (BIPV) is marketed as having 'greater kerb appeal,' certainly roof-mounted PV panels can be obtrusive but have the advantage of minimal disruption to the occupants during installation. In visual terms, the design of PV has developed little despite the hundreds of companies now manufacturing them, mostly in China, which has cornered the market due to sheer volume. If the roof covering is to be replaced anyway then BIPV becomes an option, although the lack of ventilation below results in a marginal reduction in the generating capacity, for best efficiency the panels need to be kept cool. BIPV does at least have the advantage of preventing birds and squirrels from nesting beneath the panels and enjoying the heat. Avoidance of overheating will probably see all new houses, and retrofits too, having fixed shading overhangs over windows, further increasing the potential for photovoltaics.

The government's estimate is that, on average, households that have PVs installed use 500 kWh less per year, between 40 and 60 per cent of typical household consumption, the usual assumption being that the payback time is about 10 years. Whilst the FIT was in operation, installation of rooftop solar was of obvious economic benefit to householders, now the case is not at all clear. The rate paid for solar exported energy is far less than companies charge for the supply of electricity from the mains. The likely returns are subject to a huge number of variables, the range of deals on offer, and the property's shape, orientation, slope of roof and extent of overshadowing. Ideally, the dwelling would be home-workers or retirees, in constant occupation, and able to maximise the use of their home-generated power. Otherwise, the remainder fed into the grid will own a paltry return, and in winter the panels may generate no electricity at all. So, the extent of electricity used in summer by the occupants makes a substantial difference to the likely payback time. The Microgeneration Certification Scheme (MSC) advises that an average household should be able to utilise 35–50 per cent of the generated power without a battery and that with battery storage this could increase to 80 per cent if the house is occupied for most of the day. The size of battery that would be required to meet their needs, across the seasons, would be very expensive. To be completely self-sufficient, it would need to be five to ten times larger as would the cost of installation. The UK climate is not yet suitable to going completely off-grid.

Lithium-ion batteries are improving rapidly; since 1991, they have reduced in price by 97 per cent. At the same time, their energy density has more than tripled. The energy density of natural gas remains unaltered, whilst battery capacity is improving exponentially. Currently, electric vehicles are using 80 per cent of the lithium-ion batteries being produced. Cars typically spend 90 per cent of their time immobile, around 5 per cent of the urbanised area of the United States is car park. Once electric cars are universal, car parks could install a system to interlink stationary car batteries, the stored renewable energy would help cover peaks in demand.

Once a house, and its adjacent car, can be combined to become a micropower station, PV systems will come into their own. The means to make this possible, now available in the United Kingdom (Wallbox n.d.), combines solar energy conversion that conventionally requires an inverter, with 'vehicle-to-home' bi-directional charging. This enables the exchange of electricity between the car battery, household electricity demand and the panels on the roof. Set against this optimistic outlook is the fact that the additional cobalt, copper and nickel required for making batteries will inevitably entail deep sea mining.

The Attlee government's Electricity Act of 1947 nationalised electricity companies and merged them into regional area boards. But the process was reversed as a result of privatisation by the Thatcher administration in the

1980s, which divided the market into suppliers that produce energy, retailers who sell it, and distributors who provide the infrastructure that transports it. Policy has since been effectively in stasis. Whilst decarbonisation requires a less centralised energy system, government finds it very hard to create a coherent strategy beyond a set of carbon targets. The United Kingdom has suffered 'stop start' policy measures, and an uncertain future direction towards 2050, when targets have to be achieved. Long-lasting regulations that would provide certainty have not come about. The status quo has remained dominant, and the development of localised energy is slow. In 2022, nearly two-thirds of Conservative voters wanted the utilities returned to the public sector, whilst rising prices were in danger of pushing half of UK households into fuel poverty. The 2019 Labour manifesto promised to: 'Bring rail, mail, water and energy into public ownership to end the great privatisation rip-off and save you money on your fares and bills.'

In the solar industry, there were 12,500 jobs lost in 2016 when the Conservative government wound-down the feed-in-tariff (FIT), but with characteristic flip-flop their plan, announced in April 2022, aimed to create 10,000 jobs over the next six years. Following the final closure of FIT to new applicants in March 2019, the rate at which the national area of rooftop solar was increasing had dropped significantly. Residential rooftop PV has reduced in price by 50 per cent over the last ten years, which was the motivation for cutting FIT. In 2022, the number of installations picked up again, back to about a half of the previous record year 2015. Whereas, in the past, domestic roof panels were the majority, now that 'sustainability' has become a corporate watchword, larger commercial roofs have taken the lead. FIT has been superseded by the Smart Export Guarantee (SEG) that has been in place since January 2020. Energy companies having in excess of 150,000 customers must now offer at least one SEG tariff. There are no minimum SEG tariffs, the only requirement is that the tariff must exceed zero, effectively, energy suppliers determine the level of tariff. SEG tariffs can be variable or fixed. Fixed SEG tariffs pay a given rate per kWh exported to the grid, a variable SEG tariff varies the price paid according to levels of demand. The difference between the lowest and highest rates currently on offer is considerable, the terms vary widely. In addition to SEG, energy companies are offering deals specific to each household, they are strung around with provisos, time-limited, and maybe requiring electricity, both supplied by panels and consumed from the grid, to be with the same company.

Energy companies, having no legal restrictions, are effectively free to pay whatever they wish for domestically generated electricity. When it finally ended in March 2019, FIT paid solar panel owners 33 per cent of their electricity's value, whereas energy companies on average, in 2020, offered an SEG rate of 13 per cent. In October 2022, that number fell to just 9 per cent. The Smart Export Guarantee should surely require energy companies to pay households market value for their solar power. This would encourage installation of rooftop panels, to reduce national energy prices, and cut emissions, at a time when both are dearly needed. In Birmingham, whilst FIT was still available, a 'Green New Deal' (BES 2012a) was introduced that aimed to provide private homeowners with affordable whole house retrofits, in combination with solar PV installations, using FIT as the funding mechanism. The aim was to encourage the development of local supply chains and manufacturing, but FIT was cancelled, and so was the initiative.

In Australia, using a similar market mechanism but with reliable sunshine, 30 per cent of the country's homes now have solar panels (some 2 million) whereas, in the United Kingdom, the total is still short of one million. Their feed-in tariff was only withdrawn in 2019, by which time the PV market was well developed, Australia being one of the sunniest places on the planet. The Australian version of SEG sets a minimum price as opposed to the UK version, which is something better than nothing at all. There, the best rates are paid by Tesla but this presupposes the installation of their software-controlled battery system, the capital cost of which is likely to exceed that of the rooftop solar.

Although the cost of PV panels has fallen, the total installation, including the inverter and optional battery, is likely to be double the cost of the panels. As the Energy Saving Trust puts it: 'Things are not quite that simple though. A new PV system has a life expectancy of twenty-five years, whereas over that length of time it is likely that three batteries would be required and two inverters so the final cost would be about three times that of the panels themselves' (EST 2005).

The number of suppliers has reduced in recent times, Ofgem (Ofgem n.d.) lists ten at present, although some smaller energy companies, not required to join the SEG scheme, also offer export tariffs, but the rates are generally unattractive and bear no certainty of future performance. This is clearly another case where market economics is endangering the planet. Presumably the supply companies would claim that rooftop solar only produces a meaningful return to the grid in the summer, when reduced demand makes it less valuable. In aggregate, increasing local generation will, however, allow gas-fired stations to operate for less of the time. It is apparent that as with electricity supply generally, although free to change your supplier, the individual householder has no bargaining power. Getting together with the neighbours to go off-grid is scarcely feasible at present, although there are microgrids in the United Kingdom, such as at the one at CAT in Wales. A microgrid, independent of the main grid, needs additional equipment, and the operators raise objections because part of the network will potentially be live when repairs are underway.

Although a battery is not a necessity when installing rooftop photovoltaics, a smart meter is, a further step towards an integrated system where demand and supply can be matched. A better deal for communities can be to form an ESCO, an Energy Service Company (Watts n.d.). One of the best known is Woking's Combined Heat and Power plant (Woking ESCO n.d.) but there are others that supply to local homes and businesses as is the intention of LEO (Project Low Energy Oxfordshire). With the advent of a smarter grid, more data will be available for consumers to regulate their energy consumption and thereby balance their usage in relation to supply, reducing the additional generation needed to meet high peaks in demand. This will also help to accelerate the switch over to renewables following the widespread introduction of electric cars and heat pumps. A cue could be taken from Croatia, where the 'HEPI Solar project' (HEPI n.d.) finances the installation and maintenance of small domestic panel arrays for over ten years. The investment is covered by energy savings and the surplus electricity that is fed into the grid. After the contract expires the rooftop panels become the property of the house owner.

As renewable energy becomes embedded the role of communities becomes critical. Across Europe it has been realised that combining together to form energy cooperatives, as local power suppliers, makes for strength in numbers, particularly since the current organisation of the grid around power stations will not be able to take care of future levels of demand. Localised systems based on renewables, and smart energy management, are the only way forward. EU directives on the establishment of energy communities are currently being incorporated into national legislation across the continent. Holland has the ambition to reach 50 per cent community-owned power production with 30 per cent of electricity being provided using renewables. In Italy, there are over 8000 local municipalities, over half of which have a population of less than 5000, so the potential for community energy is enormous. So, since 2019, a district of Bologna has involved the local authority, residents' associations, and the university, in the installation of solar panels on industrial rooftops. The university's involvement in this community project, known as Greta, is led by Bologna University's architecture department (Greta n.d.).

Here in the United Kingdom, in Swansea, one of Europe's largest projects involves 644 social housing properties that have been linked to communal solar PV arrays on roofs across the estate (Swansea 2022). The goal of this EU-funded scheme is for 60 per cent of the combined electricity load to be satisfied by local generation. Each participant's home is equipped with its own battery, and smart controls regulating its connection to the grid, householders are encouraged to run their larger domestic appliances during hours of sunlight for maximum benefit to their bills. This followed on from smaller pilot schemes, preparatory to the Welsh 'Mini Power Stations' project, which is to use the same principle for the retrofit of 7000 more homes in the Swansea Bay City Region. The situation in Wales is particularly challenging,

32 per cent of the nation's dwellings were constructed before 1919, and 43 per cent of its privately rented accommodation is of the same age.

Back in the days of the FIT, companies were offering 'rent-a-roof' schemes to households in return for collecting the relatively generous FIT payments, which at that time had the advantage of a guaranteed rate of return over 25 years. The occupants had the use of the electricity generated, and only the excess was exported to the grid. When FIT payments were reduced by two-thirds in 2016, rent-a-roof became less commercially attractive and by time the scheme was disbanded three years later, 'rent-a-roof' had almost disappeared. The limited number of schemes still operating offer a miserable return to householders. So at present, there is little hope for the sale of long-term rights to roof-space (cheaper than constructing a solar farm on open land) that could perhaps provide the finance for 'fabric first' street-by-street retrofit to finally happen.

The mantra 'fabric first' is to suggest that renewables are an extra to be added only after the building has been made insulated and airtight. The question is how airtight and how well insulated? LETI's graph (LETI 2021) in Appendix G, suggests that insulation pays rapid returns down to a U-value of 0.2 W/m^{2}K, after which every additional millimetre of insulation delivers diminishing returns, this is approximately the level of insulation anticipated in the 2021 revision to the building regulations. Also, the revised Part F intends that an airtightness greater than 5 m^2/m^3/hr would necessitate the installation of MVHR that will be physically impossible in many retrofitted homes. So, do we suppose that these are the limits of 'fabric first' at which point renewables become desirable, and given the de-carbonisation agenda, absolutely necessary? As with so many other aspects of retrofit these fairly basic issues have not become matters of policy, instead a variety of approaches are promoted with scant regard for practicality or outcomes.

The market-driven approach leads to unknown conclusions, how many homes will choose to be fitted with PV, how much available south-facing roof is there, and what contribution will they make to the grid after home electricity use has been deducted? For a few years, the Centre for Sustainable Energy's PlanLoCaL initiative sought to involve local communities in answering some of these questions (CSE 2013). Rather than accepting minimal recompense from energy companies, the programme drew together community groups to complete exercises that identified an area's potential for renewable energy installation. By raising awareness, the hope was that rather than the usual default reaction to obstruct change, the evident benefits would win people over. In one project, the participants were invited to markup satellite maps to show all the south-facing roofs that could be suitable for PV, distinguishing between houses, commercial and public buildings. From typical consumption figures, total community PV electricity consumption was calculated, and thereby the difference between costs and income.

PlanLoCaL recognised the need for collaborative effort within and between communities, local governments and society generally, if local renewable energy initiatives were to become embedded. Most recently, the Centre has been organising Climate Action Days for town and village councils.

A country-wide map of PV potential and a plan to mobilise local action do not exist, instead they are being left to the market. Suggested measures to kickstart the process such variable stamp duty, or council tax rates fixed according to the energy credentials of individual dwellings, have been steadfastly resisted by government. And innovations (Cousins 2021) spray-on PV (Solar Cloth/Spray on Solar 2015), for example, are being advanced around the world but are at the mercy of vested interests and cyclical capitalism, is this any way to save the planet?

References

BES. 2012. 'Go Early Pilot' (Contract between Carillion and Birmingham City Council). Birmingham Energy Savers.
BES. 2012a. 'Large Scale Green Deal Adoption – 15,000 Homes'. Birmingham Energy Savers.
Boardman, B. 2012. *Achieving Zero – Delivering Future Friendly Buildings*. Oxford: Environmental Change Institute, University of Oxford.
Cousins, S. 2015. 'Solar Cloth/Spray on Solar.' RIBA Journal, February: 28.
Cousins, S. 2021. 'Solar Eclipse.' *RIBA Journal*, March: 57.
CSE. 2013. *PlanLoCaL: Energy Efficiency and the Green Deal*. Bristol, UK: Centre for Sustainable Energy.
EST. 2005. *CE102 New and Renewable Energy Technologies for Existing Housing*. London: Energy Saving Trust.
EST. 2010. *Getting Warmer: a Field Trial of Heat Pumps*. London: Energy Saving Trust.
Greta. n.d. 'Greta Renewable Energy District'. Available online: https://projectgreta.eu/ (accessed 3 July 2022).
HEPI. n.d. 'Croatia 'HEPI Solar Project'. Available online: https://www.hep.hr/customers/household/electricity/2509 (accessed 12 September 2024).
Ingrams, S. 2017. "Do Solar Panels Affect the Value of your Home." *Which Magazine* July: 28. Available online: https://www.which.co.uk/news/article/do-solar-panels-affect-the-value-of-your-home-apHUY8c7sK6D.
LETI. 2021. 'Climate Emergency Retrofit Guide'. Available online: https://www.leti.uk/retrofit (accessed 6 June 2022).
Murray, C. 2013. 'This Poorly Thought-Out Green Deal Is a Bad Deal for Architects'. *Architects Journal*, 31 January: 5.
Ofgem. n.d. 'Smart Export Guarantee'. Available online: www.ofgem.gov.uk (accessed 18 July 2024).
Swansea. 2022. 'One of Europe's Largest Projects Involving 644 Social Housing Properties'. Available online: https://www.bbc.co.uk/news/uk-wales-63300680 (accessed n.d.).
Wallbox. n.d. Available online: https://wallbox.com/en_uk/quasar-dc-charger (accessed 23 September 2022).
Watts, C. n.d. "Energising Local Initiatives." *Sustain Magazine* 10 (102): 35.
Woking ESCO. n.d. Thameswey Sustainable Communities Ltd. Available online: https://ukdea.org.uk/thameswey (accessed 3 July 2022).

12 Measuring Success

80% Reduction and 'Retrofit for the Future'

The Retrofit for the Future Competition, finally completed in 2014, sought to find ways in which an 80 per cent carbon reduction could be achieved. The funding was £150,000 per house that included professional fees and so on, but, in many cases, the costs were greater than knocking the building down and starting again. The 86 projects involved 115 housing association properties at an average construction cost of £90,000; very few met the 80 per cent CO_2 reduction target, although useful results were achieved, including a better understanding of innovative technologies. Monitoring was carried out over two years after completion, which highlighted the lack of a consistent monitoring method, a prerequisite before embarking on a national retrofit programme. Similarly, the schemes could be initially modelled in either Extended SAP or PHPP, which aren't compatible; PHPP, although a more sophisticated program, is predicated only on the use of mechanical ventilation. Nonetheless, problems with installation at corners, junctions, edges and interfaces came to the fore, and the importance of ventilation was finally recognised.

The competition was a worthwhile experiment, but the EU Interreg project IFORE (Innovation for Renewal) that completed in 2014 spent a housing association's usual budget to achieve the same result (see part 1 case study). What it amounts to is that using standard and relatively economic measures, such as insulation, better heating systems, low-energy lighting and appliances, improved windows and doors, and sealing houses up so they are more airtight and don't leak heat, cuts carbon emissions by as much as 60 per cent. The problem is achieving the last 20 per cent; if you try using just technology, you can spend a lot of money, which is a matter of achieving the last salami slice of benefit, for example, using low-flow water appliances to achieve an extra percentage of carbon savings. The Retrofit for the Future houses employed high-tech measures such as vacuum glazing and aerogel insulation, both of which are very expensive. The cheaper way to do it is to enlist the help of the residents assisted by the 'Green Doctor.' But as of now, whilst we wait for street-by-street retrofit to begin, in deprived areas where many are on benefit and in fuel poverty, without the means to heat their homes, the choice between 'heat or eat' will require more than help from the Green Doctor.

From 2014, the Centre of Refurbishment Excellence (CORE), based in Stoke-on-Trent, was in the driving seat. In addition to organising expert conferences, CORE was to develop Retrofit Coordinator training, but the local council closed

DOI: 10.4324/9781003564997-15

the centre and programme in 2015. Training is now the responsibility of the not-for-profit Retrofit Academy. Today, a motley collection of research centres work independently, a far cry from the necessary resolve: RE:NEW a London-based project developing risk management tools is now being replaced by London Homes Energy Efficiency Programme (LHEEP); retrofit of traditional buildings is being researched by the Sustainable Traditional Buildings Alliance; and moisture movement is the concern of the UK Centre for Moisture in Buildings.

The original PAS 2030 (BSI 2019) wasn't very effective since it concentrated on individual building elements but not the junctions between one another and residents; it was revised in 2017. A clearer picture has since emerged of the various technical risks to be countered, achieving best results by accrediting qualifications, and closing the feedback loop with results from completed installations. Central to the process is the role of Retrofit Coordinator who has oversight of the entire project's progress, materials, products and details, suited one would have thought to architects but there are as yet no MArch qualifications specialising in street-by-street retrofit. This is despite the need for all PAS 2035 retrofit projects to be overseen by a qualified Retrofit Coordinator, a privileged position that architects have never enjoyed in relation to construction projects generally, at least not in the United Kingdom. The long list of capabilities required of a Retrofit Coordinator within PAS 2035 could double as an everyday architect's job description.

The complexity of the task, a reflection of the huge variation to be found between dwellings, even in a single street, initiated the RE:NEW Technical Risk Matrix, which quantifies the interactions between a long list of measures from insulation to renewables. And to promote consumer demand and protect the public interest, the Each Home Counts review contains another process flow chart, and a Retrofit Consumer Charter, TrustMark and Code of Practice, guiding the installation through to a post-completion monitoring phase and a growing data warehouse. The government-endorsed TrustMark requires installers to be members of TrustMark and to work in compliance with both PAS 2030 and 2035. Meanwhile, British Standards are finally being introduced to regulate materials, workmanship and process. So, after years of poor practice, a framework exists for best practice if one of growing complexity, the hope being that market funding will be unlocked once the technical problems have been ironed out.

Central to the endeavour is PAS 2035 (BSI 2019a), a 'whole dwelling' approach, which provides comprehensive guidance about all stages from initial consideration of building physics, through installation to completion, monitoring and evaluation; governed by another flow chart of some complexity, ventilation now having a central part to play in development of the design. It has ambitious outcomes, to avoid a repeat of the technical failures of the last ten years: Reducing energy use and the risk of overheating, guarding against water damage, whilst increasing the utility and sustainability of the building, all within a graduated risk assessment process.

Each project commences with an exhaustive assessment of its existing structure, construction and details, and identification of constraints and defects. Once again, particular attention is paid to ventilation with regard to the final level of airtightness. The PAS 2030 Retrofit Installer works to the PAS 2035 compliant design, eventually providing the householder with warranties after testing, commissioning and handover. The list of qualified participants now includes Retrofit Advisor, Assessor, Coordinator, Designer and Evaluator. Architects are deemed qualified to be adequate designers for higher-risk projects, but 'Path A' low-risk designs are down to the Retrofit Coordinator or an architectural technologist, historic and high-rise buildings are always in high risk 'Path C.'

So now the framework is in place. PAS 2030:2019 and PAS 2035:2019 were joined by publication of PAS 2031, the certification standard, in June 2019, PAS 2038 for non-domestic buildings appeared in 2021. PAS 2035 is in the process of update to promote the need for 'once-only' retrofit and away from the piecemeal measures of the past.

ECO projects now require TrustMark endorsement, as are projects funded by the Sustainable Warmth and Social Housing Decarbonisation Funds, and the hope is that the government will require other housing sectors to follow suit. But, as ever, the question is one of political will; PAS 2035 hasn't yet been brought into legislation.

Meanwhile, the welter of reports has continued unabated. In May 2016, Arup's *Towards the delivery of a national residential energy efficiency programme* (Arup 2016) pinpointed three pre-requisites for a national retrofit programme: 'Create and maintain demand at scale; Learn and disseminate the lessons from research and experience, especially from schemes that did not perform as intended; Create the right governance framework'; none of which are yet in existence.

Then in June 2016, the Energy Saving Trust's *Energy Efficiency: Recalibrating the Debate* (EST 2016), bemoaned the lack of joined-up thinking in the United Kingdom that was by then obsessing about leaving Europe, whereas in France and Germany, national plans were in place for a greener future (in Germany, the federal government has since 2007 had a system of loans, grants and tax incentives aimed at raising the standard of all pre-1984 properties across the country). But by then, all that was left of the original high intentions was ECO, to fund retrofits in social housing. The central plank of the Green Deal's 'pay-as-you-save' (PAYS), the mechanism imported from the United States to entice homeowners, had come to nothing, and the alternatives such as modified stamp duty, council tax and 'green mortgages' hadn't gained traction. The EST's general tone was one of exasperation.

In October 2017, the UK Green Building Council published *Regeneration and Retrofit* (UKGBC 2017) looking at the larger picture, how the retrofit of low-income neighbourhoods could be linked to general environmental improvements. A number of pilot larger-scale retrofit projects that had been completed in the intervening years illustrated the potential: IFORE an EU-funded collaboration between housing associations in England and France that saw the upgrading of 100 houses on the Isle of Sheppey; Warm Up Bristol a loan scheme initiated in 2014 by Bristol City Council that encouraged owner occupiers and the private rental sector to borrow over for a term of up to 15 years and pay for energy efficiencies (the scheme was wound up in September 2017 for lack of demand); Manchester Care & Repair, a charity that administers the Home Energy Plan (HELP), another loan mechanism that doesn't incur interest just administration costs. Much of the report is given over to the thorny problem of designing a market mechanism that can succeed where PAYS failed. It highlighted the many health, employment and environmental advantages that these larger-scale schemes achieved; for example, the SOAR Build social enterprise in Sheffield that employs and trains local retrofit labour and acts as a contractor and maintenance specialist. The UKGBC's hope was for two 'breakthrough innovations' that would encourage retrofit projects in low-income areas: Available funding for private homeowners, and engagement with communities. Their proposal was for the formation of Community Interest Companies (CICs), acting as a contractor, and a Local Authority Revolving Fund that would draw on a range of funding sources (unspecified), to offer new methods of retrofit financing for householders.

In March 2021, the Commons Environmental Audit Committee (GOV.UK. 2021a) regretted the lack of progress in upgrading the nation's housing stock, current measures were described as 'woefully inadequate,' the necessary skills base was still lacking, as was a credible national plan.

In April 2021, the RIBA joined the Architects Climate Action Network (ACAN), Architects Declare, Friends of the Earth, and 17 other building environment and climate campaign organisations, in calling for the proposed Future Buildings Standard to be more ambitious, including tougher carbon controls, and for the introduction of a new National Retrofit Strategy.

In March 2022, the Climate Change Committee (GOV.UK. 2022) (GOV.UK. 2019b) found key areas where there are gaps in policy, including policies needed to drive energy efficiency improvements in homes that are not fuel poor, those outside of the remit of ECO, which it described as 'currently inadequate.'

As it stands, the UK's goals are to reduce greenhouse gas emissions by 34 per cent by 2030 and 80 per cent by 2050, with net zero being met by 2050. To reach these targets, 25 million homes will need to be refurbished by 2050, amounting to two retrofits per minute or 20,000 a week. Assuming £25,000 per dwelling at a total cost of £625 billion. In addition, three million households were living in fuel poverty in 2018, made considerably worse by hikes in energy prices since, the frequently repeated choice being whether to 'heat or eat.'

The faltering attempts at setting this process underway are exemplified by the now defunct Energy Technologies Institute's *Energy Systems Modelling Environment* (ETI 2012). The study, completed in 2012, sought to accelerate the rate of retrofit by industrialising the process through design, supply and implementation, including stimulating demand by exploiting the opportunities offered by 'whole house' refurbishment. Cost effective packages, requiring least disruption to households, and an outline of the skills base required were the aims. As with so many of these studies and reports, the results on the ground are hard to find.

Low carbon Britain

Gas is planned as the main fuel for heating until 2025, though its carbon content will reduce as gas production by anaerobic digestion is introduced. After 2025, space heating demand should diminish as houses are improved, so the electricity used by lights and appliances will become an increasingly important percentage. The European Union has an ambitious programme to increase the efficiency of electrical appliances and to take less efficient models off the market. The active cooperation of building occupants will be necessary to meet the target of 50 per cent reduction in use of electricity by 2050.

The housing stock in 2050 can be expected to include more than three-quarters that have already been built, so increasing the standard of new housing will only make a dent in the challenge to meet net zero. Net zero carbon reduction by 2050 is the UK's current commitment. Domestic heating carbon emissions are to be virtually zero by then since the electricity grid is being 'de-carbonised' through the use of nuclear power and renewables. This will aid fuel security so the United Kingdom will no longer be affected by the availability of Russian gas. Ongoing is climate change; we are at present thought to be on the approach to the 4°C rise, which will put many coastal cities in the world under water. Urban pollution is increasing, and fuel costs have been rising, all of which make improving existing houses a necessity. The Future Energy Scenarios, scoped by the National Grid, call for a 75 per cent reduction in the heating and hot water energy consumed by existing homes. The Climate Change Committee has reported that without major savings in fabric efficiency the sixth carbon budget won't be met. Local community-activated retrofit is currently the only viable approach given the lack of any coherent governmental policy, whereas the European Green Deal's Renovation Wave has already settled upon a citizen-based approach to retrofit.

Retrofit history over the last ten years hasn't been encouraging; in Scotland, there have been failures with external insulation falling off schools and tower blocks. In Preston, poorly installed insulation rendered 300 houses uninhabitable due to rain ingress behind the external insulation; and most recently, and catastrophically, the external cladding implicated in the fire at Grenfell Tower.

Problem #7 The occupants

Computer simulations may predict a favourable outcome for energy-saving measures following retrofit but the few post-occupancy studies that have been completed suggest that results are actually highly variable. The multiplicity of factors – building plan type, orientation, insulation options, airtightness, and ventilation strategy – make each house an individual design problem. Though even without these constraints, the chances of achieving planned carbon reductions won't be met without the enthusiastic participation of the occupants. (Hanna, Banks-Sunan, and Robertson 2014)

> The world is littered with exemplar, low-energy buildings that don't perform as predicted...occupant behaviour can double or halve the combined effect of both the building's design and systems. The biggest single factor affecting energy use in buildings is people.
> (White 2011)

> The Performance Gap: Whilst energy models are useful to predict energy demand, we must also consider the evidence from monitoring of actual buildings in use. This data indicates that a typical building regulations standard home will require, on average, at least 60 per cent more energy for heating than is predicted.
>
> (LETI 2021)

The LETI guide is over 200 pages long, but this is, more or less, the extent of consideration given to the performance gap. The Usable Buildings Trust has pioneered understanding of the difference between modelled and actual performance and the need for post-occupancy evaluation (POE). This has progressed in relation to other building types; however, housing presents particular difficulties, not only in collecting in-use data but also in interrelating monitored figures and the sociological outcomes from user questionnaires. The causes of the performance gap are many and varied, but principal amongst them is the behaviour of the residents; neighbouring identical houses can have in excess of three times the energy consumption when comparing one to the next. A particular concern is that the cost-benefit of a national retrofit programme will be completely undermined if the extra comfort following retrofit is enjoyed, rather than the savings being realised, the so-called 'rebound effect.' Now these aspects are rising back up the agenda, section 3.3 of the draft 2025 Future Homes Standard is about building performance and concerned with the performance gap, whilst the EPC Action Plan intends that in future the calculated Heat Loss Parameter will be included within each EPC on the national database, for future comparison with the in-use figures. Clearly, though, every home in the country needs more than an EPC rating, instead an individual, phased plan for improvement which itemises the benefits. Even modest energy improvements can inhibit later upgrades; for example, replacement sash windows with sills that don't project far enough to accommodate subsequent external wall insulation. Under our current market-driven regime, random home improvement projects (usually involving attractive items such as kitchens and bathrooms with their high embodied energy) are carried out without reference to the common good. A replacement boiler doesn't attract the same appeal as visible and 'shiny' appliances, in the same way that new technologies attract the enthusiasm of politicians.

Many studies have illustrated the shortfall between predicted energy consumption and reality, which is often attributed to the impossibility of anticipating human behaviour, and the failure of designers to take into consideration different occupant activities and lifestyles. Technological data is a lot easier to gather and interpret, and standards are easier to write in relation to test results, rather than engaging with complex behavioural aspects, which have consequently taken a back seat. Although there has for some years been agreement on the need for social factors to be accorded more importance, the extent of the problem remains uncertain. Many reports have suggested that occupant behaviour might be responsible for between 25 per cent and 50 per cent of building energy use (Leaman, Stevenson, and Bordass 2010) (NHBC 2012) (Green Construction Board 2013). One concluded that changing behaviours in relation to energy could reduce gas consumption by 12 per cent and use of electricity by 7 per cent (Janda 2011). Yet another investigation, into new homes, estimated that behaviour was responsible for 51 per cent, 37 per cent and 11 per cent of measured variations in heat, electricity and water consumption, respectively (Uitdenbogerd et al. 2007). The wide variation in these results illustrates the relative simplicity of defining building archetypes and characteristics compared with the range of human behaviours (Gill et al. 2010). The necessity of addressing the issue was recognised by 'The Missing Quarter' project in Manchester, its title referred to the 25 per cent to 50 per cent energy reductions that were considered possible through behaviour change, noting that variations of up to 300 per cent have been recorded due to human choices in the home. Most certainly, after insulation and installing better-rated appliances, the journey from 60 per cent to 80 per cent carbon reduction is difficult to achieve without the enthusiastic participation of the residents.

Although these results differ markedly, what is certain is that when trying to understand and reduce domestic energy consumption, a lot of emphasis needs to be placed on human agency. Buildings and technologies may enable or constrain behaviour, but occupants generally control the levels of temperature, ventilation, lighting, and hot water, yet alone the unregulated uses of electricity. Although opinion varies, it seems clear that overall reduction in demand will only come about if energy efficiency is accompanied by measures that help consumers to reduce their usage. One related initiative is CarbonCulture (CarbonCulture n.d.), a community platform that aims to get participants in workplaces to act together towards 'the adherence loop' whereby actions are determined by shared beliefs and a common 'mental model.' 'The Missing Quarter' report, from Manchester, found that although most residents were keen to reduce the size of their bills, some were receptive to the feel-good factor of working together with friends and neighbours, and the public recognition of a contribution to civic life. Reliance on fact-based arguments is consequently likely to be unsuccessful. Rational Choice Theory, which is central to economic theory of behaviour, holds that decisions result from evaluation of the best likely outcomes in terms of maximised personal benefit, but this has proved a poor indicator of outcomes. Whilst this might be true in some situations, many other factors come into play. Research suggests that there are barriers and pre-requisites in decision-making, first the recognition that a problem exists, what to do about it and the will to seek a solution, and foresight of how individual actions might be perceived by others. The answers to these questions will be influenced by the immediate context and the viability of a particular change in behaviour, whether the views of family are friends to be considered, and the ingrained preferences that have built-up over time in terms of values, beliefs and attitudes. Given the current unprecedented rise in energy costs, energy-efficient homes can be expected to be more appreciated by mortgage companies with accompanying increase in valuations. Certainly, the levels of alarm about the world's environmental degradation are beginning to encourage changes in attitude, even amongst hard-to-reach groups such as car enthusiasts.

According to social psychologists, how the threat is interpreted is a matter of appraising the likelihood of its coming about and its probable impact, whether there is an emotional connection with the issue, and the degree of motivation in doing something about it. All of these will depend on one's particular makeup, whether a 'pioneer' who is inclined to act in pursuit of a good cause, a 'prospector' who seeks association with success and is wary of anything that puts their personal achievements at risk, or a 'settler' who is comfortable with tradition and security. Two potential influences relevant to retrofit are 'incentives' and 'norms' sometimes referred to as 'nudge' and 'think,' which combined with targeted information are likely to influence household responses. This was the basis of the Cabinet Office report *MINDSPACE* (IfG 2010), which argued that although it is difficult to change people's minds, with a well-designed strategy, the context of people's decisions can be altered. It might seem unlikely to implement societal change on the scale of a whole country, but in recent times when much of the world has accepted the need to wear face masks, collective belief might conceivably be possible. To collectively seek to limit climate change through individual action is addedly difficult in a society wedded to individual freedom, but the 'adherence loop' may rapidly change in the shrinking, globalised world of the Internet.

> … the rate at which we're exploiting the earth's resources is unsustainable: in addition to climate change, there's a rapidly accumulating list of equally daunting crises, such as capacity limits in crucial resources, deforestation, and a massive extinction of species. If the convergence of these multiple threats becomes too much to handle, our global civilisation could face a total collapse… How could this have happened? Is it an inevitable result of human nature, or is our present situation culturally driven: a product of particular structures of thought that could conceivably be repatterned?
>
> (Lent 2017, 32)

Nowadays, though, we are bombarded with information. So, decisions require 'short-cuts' that call upon innate tendencies such as the desire to seek a middle course that avoids extremes, and to err on the side of avoiding loss rather than pursuing gain, factors that make for a circumscribed form of rationality. Since humans are social animals their beliefs and actions are heavily dependent on identity within a community and social milieu, so consumer choices have a symbolic importance. In future, externally insulating a house could conceivably carry as much semiological weight as the model of car in the driveway. The importance of this aspect is borne out by research showing that some measures have limited or no impact. This is largely due to 'rebound effect,' residents taking back some potential savings in increased levels of comfort, the so-called Jevon's paradox named after the nineteenth-century economist, who noted following the introduction of steam trains that if a new technology makes more efficient use of a resource, then people use more of it. The Passivhaus Trust maintains that in extremely airtight and insulated dwellings, energy consumption is so low that the 'temperature take back factor' isn't likely to be significant. But given the economically feasible extent of retrofit in low-income neighbourhoods, subject to fuel poverty and in previously under-heated homes, 'rebound effect' is very likely, and in terms of health and comfort, desirable. One commentator even going so far as to suggest that, in terms of effectiveness, it might be better to concentrate on higher-income areas, which tend to have higher levels of consumption, and therefore increased potential for reducing demand. Set against this is the belief of Ole Fanger, whose research is the bedrock of comfort science, that humans can't entertain a loss of amenity once it has been introduced. To imagine the world without mobile phones is as inconceivable as managing without central heating.

So, the occupants truly are the 'elephant in the room.' It was rather well summed up by the Institute of Sustainability that carried out post-completion surveys at a number of Retrofit for the Future sites (IfS 2012): '...occupants often have well-established patterns of behaviour within their homes that are not easily changed. Some occupants may be less able physically or mentally to cope with the disruption and the new systems ... Models such as the Standard Assessment Procedure (SAP), based on the Building Research Establishment Domestic Energy Model (BREDEM) assume the presence of typical occupants exhibiting typical behaviour in physically well-defined dwellings. In practice, actual energy use may be higher or lower because of departures from any or all of these assumptions. Different households can consist of widely different mixes of people.'

A common issue, for example, is that smokers are inclined to open windows that will undermine airtightness and ventilation strategies, and the hoped-for energy reductions. The Institute attempted to define good practice in engagement with residents to help them through a period of adjustment, learning and adaptation. The anxiety and distress attending retrofit can be lessened by close attention given by regular visits from a technically informed respondent, ideally the architect/coordinator, who needs to be able to gauge needs that will vary with family size, ages and income. Easing householders through the upheaval of a lengthy retrofit, impacting their routines and familiar surroundings, is part of this role, which gets yet more difficult when the inevitable delays occur. The problem is, of course, that residents don't stay; soon they are likely to be replaced with different occupants, lifestyles and a similar resistance to the unfamiliar. Alternative approaches, either to make retrofit sufficiently generic for future flexibility or to build in particular requirements given that some housing association tenants remain for over 30 years, were deemed a matter of local discretion. Shouldn't that be a matter of public policy?

The behavioural aspects of retrofit are still insufficiently studied or understood, even though the occupants are the key to success or failure. The *National Energy Study*, by Sustainable Homes, published in 2014, was conducted over the heating season in 400 homes throughout England and involved 14 housing associations (Sustainable Homes 2014). It concluded that feedback on energy use resulted in savings particularly when households compared their consumption with others (as did the use emoticons expressing approval or disapproval!). Nonetheless, some households were difficult to engage with, and a lot of homes with high SAP ratings still used more than average even when weighted for the size of the home and its occupancy. Project Calebre (University of Loughborough 2013) investigated these aspects in conversation with 20 households living in 'hard-to-treat,' solid-walled, owner-occupied homes in the West Midlands. They were inclined to do work themselves rather than calling in professionals and were very resistant to moving out whilst retrofit was in progress. Although mostly concerned about comfort, original features such as distinctive single-glazed windows were prized even though they were known to be cold and draughty, the same with fireplaces. They were in favour of draughts on the grounds that the house then felt ventilated and healthy, even if it lost heat.

The 2013 *FutureFit* report published by Affinity Sutton Housing Association (Affinity Sutton 2013), analysed the outcomes of work done on 150 properties representing 22 dwelling types. A 'fabric first' approach was used resulting in savings of £557 per year for electrically heated homes but only £72 for those on gas. Whereas heating became less expensive, 60 per cent reported an increase in electricity use post-retrofit that was thought to be either due to poor weather keeping residents in doors, or because they spent the savings on additional appliances. It was the extra electricity at a higher unit cost that eroded the benefit of savings in gas usage. A lack of understanding of the new technologies, even though none of them were innovative, caused problems for residents, installers and maintenance staff. The savings made did not equate with predictions from SAP calculations perhaps because some tenants were underheating their homes and because some had difficulty operating heating controls. Sixteen per cent of technological problems were found to be due to incorrect installation. This shortfall in knowledge as to how occupants interact with technology, results in a lack of explanation as to how behaviour determines energy consumption, commonplace across projects is the extent to which variations are due to difficulty in understanding heating controls. The national trend from the 1970s onwards was for a progressive fall in energy costs relative to household income but in recent times this has come to a resounding halt.

Given the recent extreme rises in energy costs, simple messages about turning the thermostat down or running washing machines at lower temperature are becoming embedded, whilst fuel poverty is an ever-growing acute problem. Costs could be made more evident with more informative and intuitive controls and accurate bills, the widespread availability of real-time displays of electricity use, and technology to turn-off non-essential electrical devices.

Preparatory to the ill-fated Green Deal, a number of resident engagement programmes were carried out, adding to the volume of academic research that has been given over to the topic. These were usually by housing associations, at a quite small scale, far smaller than would be statistically valid, a problem that besets many aspects of retrofit. Social housing providers having made energy efficiency improvements are anxious to see the expenditure reflected in better performance, including that of their tenants. So, in most cases, the methods adopted are similar, distributing free-issue energy-efficient consumables such as light bulbs and enlisting the help of children as 'energy detectives,' to check on and enlighten their parents. One of the most interesting of these was Relish™ 'Residents 4 Low Impact Sustainable Homes' carried out by Worthing Homes (Worthing Homes 2009), to educate their residents on the impact lifestyle can have on consumption and bills. Some were given tuition but no retrofit works, and others were given both or the home improvements only. The initial pilot involved just four households; one that received both achieved a reduction of 29 per cent; whereas without the educational input, another family managed only 4 per cent; compared with a high energy-use household that received education only that saved 18 per cent; and a low energy one that actually increased by 2 per cent. Some seven months after the pilot ended, the first household continued to make savings whereas the others fared less well.

It was concluded from the first phase of Relish™ that the residents were receptive to changes in behaviour relating to electricity use for lights and appliances, but less so when it came to living with lower room temperatures and taking showers rather than baths. Plans had to be flexible because some were solely interested in reducing their bills, but others hoped to make a positive environmental contribution. The way to better engagement was to invest time early on to encourage better participation. What makes Relish™ particularly interesting for the national retrofit effort is that its ethos is completely different from the technical approach taken by EnerPHit, for example. The underlying aim was to achieve the maximum with the least input of manpower and materials, consistent with the usual budget of housing associations, and able to be carried out with residents continuing living in their homes. Each house was evaluated for the best retrofit specification from a list of fairly basic measures, including cavity insulation, draught-stripping, loft insulation and boiler controls. It showed that although an 80 per cent reduction wasn't achievable, the outcomes were not far off; the residents modified behaviours contributing 20 per cent of the total. Now that energy costs have risen stratospherically, rather than the reluctance to participate that was evident at the beginning of the project, retrofit proponents might be pushing at an open door. A final outcome was however a technical innovation, Relish Smartwire, a similar concept to that found in hotels, but instead of a keycard that turns on the lights, the Relish homes have a switch by the front door, and separate power outlets for electrical equipment that doesn't need to be kept on standby when the property is empty. Increasing UK electrical demand has been closely associated with the increasing number of appliances and their reducing purchase cost. All of the four first-phase households had at that time, televisions in every bedroom. Relish Smartwire separates those systems, such as smoke detectors, that need to be on all the while, from those, such as televisions, that would otherwise be constantly plugged-in and consuming electricity. Standby power, according to the Energy Saving Trust is responsible for 9–16 per cent of electricity costs.

Relish™ was an exceptionally well-designed and intensive project. All the members of the team were trained beforehand, and most of the interactions were on a one-to-one basis. The residents were given a handbook on completion of the works and an individual energy plan for the next six months. Resident 'energy champions' were recruited to inspire their neighbours, there were workshops, a poster campaign with engaging graphics, a demonstration show-flat used as an information hub, all the participants were visited every two months throughout for review of their ongoing energy plans. Resources deriving from Relish™ were made available to other housing associations, such as Southway Housing Trust in Manchester which used the Residents Energy Guide for a larger-scale education programme. They similarly recruited 'champions' from amongst their tenants, instituted a training programme for their staff, drop-in surgeries, home energy audits, distribution of energy monitors and energy-saving lightbulbs, and three 'eco-demonstration' homes. So, the techniques of housing association tenant engagement projects have become well established, perhaps paternalistic and with uncertain results but enjoying the benefits of a caring relationship between the provider and residents. The rest of the stock, by far the largest proportion, homeowners and the private rental sector, are a very different story and how to address them remains a mystery.

It's true that there is a vast literature from sociology, psychology and anthropology that has been used by advertising to influence the public, which has been classified into 'market segments,' and their purchasing preferences. Pure economic theory that individuals act in order to achieve the greatest benefit for themselves is undermined by the facts. Because humans are social animals, our decisions are often made as a member of a group, whether family or community. So, there is a strong disincentive from stepping outside what the group would regard as 'the norm,' what within that context is regarded as normal behaviour. Similarly, achieving status within the group can be an influence, but with the introduction of new technologies norms can change and spread through social channels of communication. For example, there is reciprocity between contemporary work patterns, the use of cars and availability of supermarkets, and household ways of eating and living. Some attitudes are inherited from childhood and are immutable, such as baby boomers who continue to 'scrimp and save' even though that may not be necessary given their current circumstances. Consequently, behaviour change with its technical, social and economic components is complex and hard to achieve. Habitual behaviours, such as turning lights off when leaving a room, are more difficult to address than conscious decisions. One-to-one approaches that seek to determine the particular outlook of each household tend to be more successful than factual leaflets, particularly if the intermediary architect/coordinator has the necessary rapport. Social identity, values and worldviews, the self-perception of individuals derived from circumstances of their upbringing, locality and education, can influence an enthusiasm for energy-saving initiatives. The Green Deal that was wound up, after two years, in 2015, managed to attract only 14,000 households to take out loans within the scheme. Whereas there is a flourishing market for double glazing even though this is a less effective energy-saving measure than others that the Green Deal supported. Consumer preference tends towards measures that are achievable with not too much inconvenience, relating to comfort, enhancing the homes appearance, and its role as a status symbol; factors that need to be better understood.

> For businesses and consumers together, we can develop, promote and support brand values associated with low energy use and longer-lasting embodied energy. For individuals, we need a small social revolution, and at present we're particularly short of pioneers who can demonstrate that they, at least, are unafraid of the challenge.
> (Attwood and Cullen 2011) (Attwood 2012)

PlanLoCaL's *Getting People Involved* (PlanLoCaL 2013) sets out a methodology for researching the characteristics of people in a particular community and directing social marketing projects to address barriers to behaviour change. These subtleties include invoking an 'injunctive norm' that even if the desired outcome hasn't yet become the local norm, can promote the desirability of change and introduce transformative comparison, and competition, within the group. The strength of prevailing norms will vary according to the degree of social cohesion and diversity of the neighbourhood mix. Nonetheless, the programme won't achieve success unless the outcome is 'self-efficacy' whereby people are able to believe that change is possible and likely to be effective. A variety of techniques have been tried to 'nudge' the natural conservatism of humans and to make the unfamiliar familiar through small changes, which is where the show houses and freebee light bulbs enter the picture. The handouts play on the fact that gifts are perceived as a debt that needs to be repaid by reciprocal actions. Underlying these stealth tactics is the previous lack of success in convincing the public at large about the dangers of climate change, until now it has been impossible to imagine how everyday life will be impacted: '…fear can only motivate behaviour change if a personal vulnerability is felt.' These techniques presuppose a targeted programme with dedicated staff, and locally appointed champions, possible within the defined remit of housing associations. On a larger scale, they will have their uses, but a meaningful social profile may prove difficult to find in the extremely diverse and ill-defined neighbourhoods of inner cities. We are, however, approaching the point when a general understanding of the climate crisis will result in both fear and belief. During World War II, newspapers contained only news of the war because that was the crisis, does that point have to be reached before nationwide retrofit becomes an achievable imperative?

> There is a danger that government will fall into the trap that, in the hope of delivering rapid results on a large scale, they will promote initiatives that work on paper but not in practice. They may favour things that make profitable businesses but do not serve the public interest… creating unnecessarily expensive, complicated and technically risky approaches which building evaluation work predicts are unlikely to work well or prove good value in practice. In UK housing, there has also been too much focus on heating and renewable energy, whilst burgeoning electricity use, improved building procurement methods, and opportunities for understanding and influencing behaviour have been largely overlooked.
>
> (Leaman et al. 2010)

References

Affinity Sutton. 2013. *Affinity Sutton Housing Association FutureFit Report*. Available online: https://assets.publishing.service.gov.uk/government/uploads/system/uploads/attachment_data/file/656866/BEIS_Update_of_Domestic_Cost_Assumptions_031017.pdf. (accessed 2 March 2022).

Arup. 2016. *Towards the Delivery of a National Residential Energy Efficiency Programme.* Available online: file:///C:/Users/User/Downloads/residentialretrofit2016-3.pdf. (accessed 9 May 2022).

Attwood, J. 2012. 'Questing beast of energy policy'. *Cambridge Alumni Magazine issue 65.*

Attwood, J., and J. Cullen. 2011. *Sustainable Materials – With Both Eyes Open: Future Buildings, Vehicles, Products and Equipment – Made Efficiently and Made With Less New Material (without the Hot Air).* Cambridge UK: UIT.

BSI. 2019. *PAS 2030: Publicly Available Specification.* Available online: https://www.bsigroup.com/en-GB/standards/pas-2035-2030/ (accessed 13 July 2022).

BSI. 2019a. *PAS 2035: Publicly Available Specification.* Available online: https://www.bsigroup.com/en-GB/standards/pas-2035-2030/ (accessed 13 July 2022).

CarbonCulture. n.d. Available online: https://theodi.org/organisation/carbon-culture/ (accessed 7 August 2022).

EST. 2016. *Energy Efficiency: Recalibrating the Debate.* London: Energy Saving Trust. Available online: https://energysavingtrust.org.uk/report/energy-efficiency-recalibrating-the-debate/ (accessed 20 June 2022).

ETI. 2012. *Modelling the UK Energy System: Practical Insights for Technology Development and Policy Making*, Loughborough UK: Energy Technologies Institute. Available online: https://www.eti.co.uk/library/modelling-the-uk-energy-system-practical-insights-for-technology-development-and-policy-making-2/

Gill, Z, M. Tierney, I. Pegg, and N. Allan. 2010. "Low-Energy Dwellings: The Contribution of Behaviours to Actual Performance." *Building Research and Information* 38 (5): 491–508.

GOV.UK. 2019b. *UK Housing: Fit for the Future*, Climate Change Committee. Available online: https://www.theccc.org.uk/wp-content/uploads/2019/02/UK-housing-Fit-for-the-future-CCC-2019.pdf. (accessed 10 February 2022).

GOV.UK. 2021a. *Energy Efficiency of Existing Homes,* Commons Environmental Audit Committee, March. Available online: https://publications.parliament.uk/pa/cm5801/cmselect/cmenvaud/346/34604.htm (accessed 15 September 2022).

GOV.UK. 2022. *Progress Snapshot, Climate Change Committee, June.* Available online: https://www.theccc.org.uk/uk-action-on-climate-change/progress-snapshot/ (accessed 15 September 2022).

Green Construction Board. 2013. *The Performance Gap: Causes and Solutions.* London: Green Construction Board.

Hanna, P, L. Banks-Sunan, and S. Robertson. 2014. 'Innovation for Renewal: A Report on the UK Behavioural Change Project', unpublished.

IfG. 2010. *MINDSPACE: Influencing Behaviour through Public Policy.* London: Institute for Government'. Available online: https://www.instituteforgovernment.org.uk/sites/default/files/publications/MINDSPACE.pdf. (accessed 21 February 2022).

IfS. 2012. *Low Carbon Domestic Retrofit 'Occupant-Centred Retrofit: Engagement and Communication – Analysis of a Selection of Retrofit for the Future Projects.* London: Institute for Sustainability.

Janda, K. 2011. 'Buildings Don't Use Energy: People Do." *Architectural Science Review* 54: 15–22.

Leaman, A, F. Stevenson, and B. Bordass. 2010. "Building Evaluation: Practice and Principles." *Building Research and Information* 38 (5): 564–577.

Lent, J. 2017. *The Patterning Instinct: A Cultural History of Humanity's Search for Meaning.* New York, NY: Prometheus.

LETI. 2021. *Climate Emergency Retrofit Guide.* Available online: https://www.leti.uk/retrofit (accessed 6 June 2022).

NHBC. 2012. *Low and Zero Carbon Homes: Understanding the Performance Challenge.* London: NHBC Foundation.

PlanLoCaL. 2013. *Getting People Involved.* Available online: https://www.greenopenhomes.net/downloads/file/CSE_PlanLoCaL_overview.pdf (accessed 12 September 2024).

Sustainable Homes. 2014. *National Energy Study*. Available online: http://www.arch-housing.org.uk/media/82183/nes2.pdf. (accessed 20 July 2022).

Uitdenbogerd, D, C. Egmond, K. Jonkers, and G. Kok. 2007. *'Energy-related Intervention Success Factors: A Literature Review'*, ECEEE Summer Study Proceedings, France.

UKGBC. 2017. *Regeneration and Retrofit*. London: UK Green Building Council. Available online: https://www.ukgbc.org/ukgbc-work/retrofit-led-regeneration/ (accessed 10 July 2022).

University of Loughborough. 2013. 'CALEBRE – Consumer Appealing Low Energy Technologies for Building Retrofitting', September. Available online: https://www.cibse.org/media/cw4nmpyi/calebre-cibse-presn-v3-jul13-d-loveday.pdf. (accessed 9 December 2021).

White, C. 2011. "Kick the Habit." RIBA Journal October: 67–68.

Worthing Homes. 2009. *Relish*™ *'Residents 4 Low Impact Sustainable Homes*. Available online: https://www.worthing-homes.org.uk/relish-phase-2(accessed 12 September 2024).

13 The Progress of Retrofit

Much of the pioneering development of retrofit has so far been carried out within social housing by housing associations, organisations that in many instances represent a last outpost of social welfare. Through a series of previous programmes, Decent Homes, CESP and CERT, housing associations have mechanisms ready for the next stage of deeper retrofit, maybe funded by ECO4 or the support available in London, and a culture of resident representation and community engagement.

The ongoing difficulties are only too evident in the 2022 report by the Local Government Association, *Hard to Decarbonise Social Homes* (LGA 2022), yet another modelling exercise. This one used Parity Projects 'Portfolio' Stock Assessment Tool, in association with data from the 2016 English Housing Survey. The focus on this occasion was the anticipated move away from mains gas to heat pumps, which, whilst achieving carbon reductions, was going to be unaffordable for social housing tenants, without the fabric of their homes having been improved. Given the rapidly increasing cost of achieving the highest standards of carbon reduction, 90 kWh/m^2 was set as the target for the Social Housing Decarbonisation Fund, because to drive down from 90 kWh/m^2 to 50 kWh/m^2 was estimated to require ten times the investment. Allowance was made for the energy characteristics of different dwelling types, according to their location and variations in regional climate. Of the types that failed to meet the 90 kWh/m^2 goal, bungalows and detached houses were at the top of the list. A notable problem was the restriction placed by planning authorities on external insulation and solar panels, particularly in conservation areas. The lack of a national register, a Domesday Book, means that many local authorities do not know, for example, which of their properties are listed. In addition, the restrictions placed on energy improvements in conservation areas are subject to local fiat. One solution put forward for these old and hard to decarbonise homes, other than demolition, was to sell them on into the private sector where legislation is that much laxer, even though that would reduce the stock of public housing and do nothing for the nation's carbon reduction undertakings. All of which illustrates how far, or how little, progress has been made over the years towards a rational UK approach to climate change.

Retrofitting the owner-occupied stock, two-thirds of the total, is proving particularly difficult. Older properties are a large proportion of this sector, added to which owner-occupied houses tend to be bigger and are more often detached,

DOI: 10.4324/9781003564997-16

both of which add to the cost and complication involved in energy improvements. The current but very partial approaches to this problem are ECO+ for insulation work, and the Homes Upgrade Grant, which allows private landlords to apply to a participating local authority for two-thirds funding of energy upgrade schemes.

Overall, it is evident that the limits of simple energy-saving measures are close to being met, but the country is a long way from achieving the Passivhaus level of performance that Brenda Boardman advocated back in 2005. The rollout of a national retrofit programme remains as remote as ever, street-by-street implementation is still a rarity; one-off demonstration projects remain the only exemplars. So, in March 2021, the UK Government's latest undertaking was published, a mere 50,000 low-income and social housing properties to be fitted with wall, floor and roof insulation, but only if the local authority has been able to secure access to The Social Housing Decarbonisation Fund. Their most recent emphasis, the replacement of gas boilers with heat pumps, is covered by the Boiler Upgrade Scheme launched in 2022, and some help for homeowners may be provided through LA Flex. Yet we still have no clearly stated ambitions for the standard of retrofit to be the general benchmark, who will monitor the results, and how will it be established that success has been achieved.

The United Kingdom is far from achieving one retrofit per hour. Social housing retrofit schemes continue but that is all at present, apart from enthusiastic private sector 'first adopters,' as was all too evident at the Building Centre's Retrofit 23 exhibition. New build is absorbing all the available labour, not least because retrofit is complicated and entails working with occupants. On-site expertise needs to be developed in order to arrive at the standards expected of boiler installation, for example, and there are problems inherent within the industry; even in the same organisation, there is often a divide between designers and installers.

Costs remain a problem, to achieve a 60 per cent CO_2 reduction had been thought to be achievable for between £25,000 and £50,000 if efficiencies could be achieved at scale, but building construction is not at present at all like the mass production of cars or planes. Community-based retrofit programmes can hopefully happen under the auspices of PAS 2035 but grant support will be needed for poor households or loan support for the others. A clue may be the approach taken in Germany, where money is available from a dedicated bank that funds retrofit at a lower interest rate, but each project needs to be signed off by a qualified person to release the cash.

Retrofit and performance

Discussion about retrofit inevitably starts with statistical hand-wringing about the extent of the climate change problem and the response required in terms of houses to be retrofitted. The debate then becomes mired in the extreme economic, policy and practical difficulties of realising the scale of programme that is necessary. This discussion excludes wider issues about the impact of retrofit on cities, the advisability of some of the techniques involved, its impact on communities, and how this programme will be viewed in the future, given that climate change will create a different context for the environment of buildings.

> ...architecture and urbanism over the later part of the twentieth century, relentlessly cut away the social and environmental foundations in which the Modern Movement and its nineteenth-century predecessors had been anchored – foundations that had allowed architecture, in turn, to act as a stabilizing and socially embedding element within society. Instead, whipped up by the boom in global capitalism and the spread of the culture of the spectacle, architects across the world launched into a new, aggressively individualistic world view, oriented towards ephemeral display.
> (Glendinning 2010, 167)

For 'starchitects' to embrace eco-design would be to step down from the heights of Parnassus to the lowly and alien world of green architecture, such is the divide between the two cultures within current architecture.

> We always hear that green architecture 'looks bad' and most of it does. At the end of the day, though, separating your trash is probably a greater contribution to world architecture than Bilbao.
> (Sorkin 2011, 6)

Retrofit with its seemingly prosaic enthusiasms occupies a yet lower level of enterprise. 'The fetish for form that has characterized the profession and the schools for the last few decades has slighted much more urgent matters...issues of the environment and social justice (linked inextricably) are those I feel must foreground both the ideology and the pedagogy of contemporary architecture' (Sorkin 2011, 361).

'High' design continues as ever in the footsteps of high Modernism. Even concrete, so long derided by the public, is back in the pantheon along with an ever-continuing pursuit of 'transparency.' Extensive glass and concrete do not suggest frugal consumption of either embedded or operational energy, but these buildings, and their designers, continue to win accolades.[1]

> Adaptive use is the destiny of most buildings, but the subject is not taught in architectural schools. Any kind of remodelling skills are avoided in the schools because they seem so unheroic.
> (Brand 1994, 109)

Meanwhile green design has assumed the status of folk art, whilst retrofit is perceived by the majority of architects as maintenance writ large, a job for surveyors and engineers. So perhaps, the term 'retrofit' should have been used throughout this book interchangeably with less heavily loaded words like refurbishment or rehabilitation. Or perhaps not: 'Refurbishment sound too dull. Retrofit sounds better, but doesn't really cut it either, as it implies a lot of insulation, double glazing, airtightness tapes and some updated technology thrown in for good measure' (White 2011).

Architects as retrofit leaders

The coincidence of the 2008 recession with mounting worries about climate change and likely soaring energy costs in the future has also fuelled a growing hostility to iconic architecture by advocates of sustainability... The effects of any radical change in the economics of building must inexorably feed through in due course into patronage patterns and pressures, and then onwards (via the semi-detached relationship of architecture and society) into architectural ideologies and architectural forms.

(Glendinning 2010, 139)

The economic crash of 2008 was thought to have changed the agenda, shifting attention away from shiny new buildings towards the existing building stock, but unfortunately, for street upon street of existing housing, this has not proved to be the case.

Mercifully, I sense that amongst the younger generation of architects there is a reaction similar to that of Nauman who, at the birth of the twentieth century, said 'we have had enough of the extraordinary: what we need is the self-evident'. A phrase that I often hear now is the pursuit of 'the poetics of the ordinary'... the original emancipatory commitment of the Modern Movement in its aim for the engagement and participation of 'the man in the street'.

(Wilson 2007, 19)

Now, retrofit is a national priority, and the growth of the 'green economy' is seen as the 'next big thing.' Economic benefits can accompany environmental benefits and the creation of jobs within a new area of enterprise. There are however many obstacles to overcome, householders aren't aware of the opportunities and benefits, architects are presented with a maze of options and new untried products are in abundance. As yet there are few competent installers, contractors are unacquainted with new approaches, and local authorities have trouble approving unfamiliar proposals. In short, retrofit needs an integrated approach led by a lead consultant with the ability to grasp and direct the whole operation, a role heaven-made for architects searching for the future, and a new aesthetic.

Architectural skills required?

New theory must make detailed surveys of what architecture has discarded in the wholesale handing over of everything remotely scientific and quantitative to consulting engineers...these are activities that do not currently form the stuff of architectural practice (let alone produce design fame), and they

are taught hurriedly (if at all) by the least design-adept teachers at school…
This is about raising a submerged Atlantis of architectural sensibility, a realm of facts and insights that can support popular connoisseurship of the qualities of building equal to that devoted to the valuation of music, cars, and movies. To take this material seriously, technically and poetically, will help us to make a powerful case that architecture matters at all and can produce genuine effects that people will notice, appreciate, measure, value and ultimately demand.

(Benedikt 2005, 20)

Potentially, the UK's retrofit programme could present an enormous £500 billion opportunity to kick-start the country's economy and revitalise the practice of architecture. This figure blithely assumes that the 20 million homes to be improved will do so at an average cost of £25,000 per house to achieve 60 per cent carbon reduction. The additional 20 per cent, towards the overall goal of 80 per cent reduction by 2050, is to be achieved by 'decarbonisation' of the electricity grid, and rollout of local heating and renewable energy systems. This amounts to half a million dwellings retrofitted every year, or one project completed per minute.

PAS 2035 envisages that in addition to Retrofit Coordinators (who should be architects), low-carbon surveyors, assessors and advisors will be involved throughout the retrofit process. Only accredited advisors will be qualified to carry out a survey – an enhanced Energy Performance Certificate survey – with recommendations tailored to the requirements of particular households. The outcomes will result from the advice given and agreement about the work to be done.

At the launch of the Green Deal in 2012, the Architects' Journal (Murray 2013) (Fulcher and Mark 2013) bemoaned the way architects were being painted out of the retrofit scene, and this was one indicator of the Green Deal's inadequacies. Architects are the most trusted for sustainable holistic design, opined the magazine, but would be supplanted by Green Deal Advisors, who, with a mere three days of training, were free to advise on all aspects of retrofit. Not one architect signed-up to be involved in the development of the Green Deal. The role of the advisor was contrary to what all architects know that only a design-led approach ensures a healthy as well as energy-efficient environment. Ten years on, nothing has changed, but the need for architects to 'seize the reins' has become urgent.

Note

1 'Now, super-clients engage super-architects to design self-promoting buildings whose frantic forms and bright shiny materials can destroy the cohesion of street frontages and public spaces: individualistic buildings for atomised communities, architecture without a conscience'. From Berman, A. 2015. "Seldom do Architects Ask Simple Questions." *Architects Journal* 11 December: 49.

References

Benedikt, M. 2005. "Less for Less Yet: On Architecture's Value(s) in the Marketplace." In *Commodification and Spectacle in Architecture*, edited by W.S. Saunders. Minneapolis, MN: University of Minnesota Press.

Brand, S. 1994. *How Buildings Learn: What Happens after They're Built*. New York, NY: Viking.

Fulcher, M., and L. Mark. 2013. 'Green Deal Branded Bad for Built Environment and for Profession.' *Architects Journal*, 31 January: 13.

Glendinning, M. 2010. *Architecture's Evil Empire: The Triumph and Tragedy of Global Modernism*. London: Reaktion Books.

LGA. 2022. *Hard to Decarbonise Social Homes*. London: Local Government Association. Available online: https://www.local.gov.uk/publications/hard-decarbonise-social-homes (accessed 10 June 2022).

Murray, C. 2013. 'This Poorly Thought-Out Green Deal is a Bad Deal for Architects.' *Architects Journal*, 31 January: 5.

Sorkin, M. 2011. *All Over the Map: Writings on Buildings and Cities*. New York, NY: Verso.

White, C. 2011. 'Kick the Habit.' *RIBA Journal* October: 67–68.

Wilson, C. St J. 2007. *The Other Tradition of Modern Architecture: The Uncompleted Project*. London: Black Dog Publishing.

Part III
Towards a New Utopia

14 The Art of the Imagination

Whilst politicians still promote growth that is increasingly difficult to find, the failed record of sustainability has given place to 'regeneration.' Back in 2016, Yuval Noah Harari's *Homo Deus* (Harari 2015) predicted the rise of new 'techno-religions,' his focus was on artificial intelligence and bioscience, but now the state of the world's climate and environment is elevated in concern to one of clear and present danger. What precisely a Gaia faith might be remains to be seen, but in an increasingly interconnected Internet diaspora where everyone found themselves wearing face masks, and are now turning down the heating, the seeds may already be sown. Environmentalists speak of attention, care and consciousness, their concern for the spiritual dimension of the Anthropocene, whilst increasing activism is accompanying the alarming incidence of climate-related events.

> Which construction of meaning will ultimately triumph and replace our current structure? This is perhaps the most crucial question facing humanity because whichever one triumphs may ultimately control the future direction of the human race. And it's a question that we, the current generation, will all participate in answering through the choices we make.
>
> (Lent 2017, 436)

Now we are so far into the Anthropocene that the realisation of 'safe' global warming targets seems unlikely. Carbon dioxide emissions keep rising inexorably as does the level of deep uncertainty about unforeseen tipping points, the Thwaites Glacier in Antarctica, methane from melting tundra. Clearly, the world's hope for rescue from the worst of outcomes cannot be faith in as yet unavailable technologies. Or indeed those already available, electric cars still need the polluting manufacture of steel, let alone mining the seabed for lithium. Yet the solution is clearly in our own hands, the fiction that is capitalist perpetual growth, the belief system that now dominates the world, needs a reset similar to the 'swerve' that issued in the Renaissance or Luther's theses nailed to Wittenberg's church door.

Philosophers sometimes like to speak of 'the Event' – a political revolution, a scientific discovery, an artistic masterpiece – that is, a breakthrough which reveals aspects of reality that had previously been unimaginable but, once seen, can never be unseen.... Societies around the world appear to be cascading towards such a point.

(Graeber and Wengrow 2021)

Shared beliefs are the essential leitmotif of any society, and these are capable of radical realignment, as saw the eighteenth-century ascendancy of humanism over religion. As a recent paper in Climate Change (Hinkel et al. 2020) noted: 'Narratives are socially constructed 'stories' that make sense of events, thereby lending direction to human action.' An endeavour, to attain a future for humanity which is happier, inclusive without inequality, limiting climate and environmental damage, is a utopia both to be hoped for and which can become the credo for retrofit architects. The Modernist utopia that once held the profession in its spell (Tafuri 1976) needs in a de-growth world to be superseded by a new set of beliefs.

'The world of reality has its bounds. The world of the imagination is boundless.'
Jean-Jacques Rousseau

In Plato's cosmology, our conceptual consciousness, expressed through reason, linked us to the divine. The early Christians transformed this into the conception of an immortal soul existing, after the body's death, with God in heaven. Descartes reformulated this dualistic framework into the modern, scientifically acceptable mind-body split, identifying the human capacity for thought as the essence of our existence.

(Lent 2017, 424)

It was in opposition to this rationalist, scientistic world view, the smoking chimneys of industrial England, that the Romantics found a different path:

Wordsworth and Coleridge walking in Somerset in 1797 wanted to change the world - ridding the mind of all that enlightenment thinking that clarity and rationality and analysis is the way to understand something, actually the way to understand is being physically and bodily in the world.

This was a radical shift in consciousness, a reset of our place in the world, this deep shift in the relationship to nature is a shift away from thinking that we are outside of nature, towards a new view that we are deeply embedded in it.

If you only talk in an instrumental way about environment and resources, what you can use and what you can conserve, you are missing the essential point that we are not separate from it, we are part of it, it flows through us, there is no boundary at the skin. Wordsworth and Coleridge saw poetry as a way of changing the world, poetry dispenses with dead forms of thinking, at the root of dead societies, dead politics, dead human relations. If you can get true communication happening through the poetic you will flow to a better relationship between people, a better understanding of your role in nature. Nowadays, an incredibly unfashionable idea but nonetheless true. Adam Nicolson interviewed on the BBC *Today* programme 2019 (n.d.) and (Nicolson 2019) (Nicolson 2020).

Recent pointers such as the establishment of the Anthropocene Architecture School, the publication in the United States of 'Out of Architecture' an uncomplimentary dissection of the profession and its education, and the 2021 anonymous posting on Twitter of '*An Architecture Degrowth Manifesto*' confirm the state of architecture to be, as with the climate, one of crisis.

It was in an attempt to free culture from what Henry James called this 'overburdening sense of the past' that Modernist architects, formed by Futurism, attempted to erase its traces from their architecture. This urge to escape history was joined to a therapeutic program, dedicated to the erasure of nineteenth century squalor in all its forms, that proposed an alliance between the hygienists and the architects that would be reinforced on every level by design.

(Vidler 1992, 63)

The Modernist ethic one of erasure and pristine replacement is a reductive formula that ignores those intangible qualities of mood, remembrance, homeliness, not readily measured or quantified, but which characterise cities as a lived experience. Can city-scale retrofit, through an empathetic understanding of localities and their residents find a new way forward – a new architectural project, a new utopia?

Retrofit realigned

Discussion about retrofit has reflected the government's view that as a means of achieving the urgent requirement to reduce carbon emissions, and a way of reducing energy demand and fossil fuel dependence, it is a relatively simple 'technical fix.' This purely technical discussion is at the level of instrumentality, carbon saving is the goal, and the complex arguments are reserved for tricky questions of logistics, implementation and funding. But a one-dimensional programme of national retrofit, directed only to reducing emissions, risks unforeseen side effects.

Those with long memories will recognise the lessons of a previous UK housing programme, driven predominantly by speed of construction and cost, concrete panel system flats, a venture that ended with Ronan Point and the subsequent demolition of mildewed blocks across the country. The uprooting of communities and its social consequences were the unconsidered side effects.

An architectural view, concerned with not just energy saving but the quality of environment, not only aesthetic, but to do with social welfare and quality of existence, begs questions about internal and external qualities of place and space post-retrofit, the impact on residents and communities, and their active engagement in the process.

Retrofit and utopia

Utopia is being revisited, and from a number of different directions:

> We can, individually and collectively, align ourselves with that deep inner yearning in all living things – for growth in cyclical complexity, toward a fuller, more intertwined and inclusive self-realization. Enlivenment economics view our goal as leading a fuller life, writes philosopher Andreas Weber: "If we adopt this perspective, we will begin to see that something is sustainable if it enables more life – for myself, for other human individuals involved, for the ecosystem".
>
> (Stoknes 2021, 293)

It has become commonplace to deride utopianism as a story of inevitable disappointment, but it was from the initial utopian impulse, which gave rise to the workers' settlements built in Germany in the 1920s, that all the various and subsequent strands of Modernism derived, followed by the large second-generation architectural waves that dwarf the diminished ripples of our own time. Utopianism is enjoying a revival; Greta Thunberg is a utopian: 'Climate change is not only a threat, it is above all an opportunity to create a healthier, greener, and cleaner planet which will benefit all of us' but for architects the context is very different than at Modernism's beginnings. Then, anxious to cast aside the bric-a-brac of the old-world order, the city became the tabula rasa onto which a master plan was imposed. Now, in the age of retrofit, the city becomes an ecosystem to be understood and cared-for, part of nature not distinct from it; a process without a clear visual conclusion, one that is moulded within communities, a mindset that is distinctly feminine, rather than that of jostling, male, star-architects flaunting their icons. For architecture, this then is a matter of both means and ends.

The first generation of modern architects, and their utopianism, an attempt to eradicate architecture's history of association with power and authority, was seen even some 50 years ago as having failed. Theirs was an inversion of the nineteenth-century notion that the artistic merit of cultural artefacts is dependent on the spiritual and moral well-being of society, their architecture rather than coming out of society was to be a force for change in society, by the imposition of order on the post-World War I seeming chaos:

> The artist reads the spiritual parallels of his time and represents them in pure form. When such spiritual common property is lacking, there is nothing left for him but to build up his metaphysical element from his own inner resources … If the signs are correct, the first indications of a new unity which will follow the chaos are already to be seen. Walter Gropius.

Theirs was in fact a dual utopia, on the one hand an architectural proposition that was the inversion of all that went before, objects stood in cleared landscapes, transparency and light replaced the dark and dank Victorian streets. Allied to which, the way would be clear for social and political revolution, replacing the hidebound superstitions of the past with the offer of a new autonomous way of life for the fit, agile and free citizenry of the future. This radical programme has been resisted not only by the populous (though nowadays in our atomised, gig-economy many lives are lived out in the new breed of high-rise) but also by the organic ecosystem of the city itself.

> The idea of a 'modern' architecture, at least insofar as it was consciously identified with the idea of the avant-garde, held two dominant themes in precarious balance. The one, stemming from the demand for cultural revolution and a sense of the exhaustion of traditional academic forms, stressed the need to remake the language of the art, to explode the conventions and of the debris to construct a way of speaking adequate to the modern movement. The other, more tied to the tradition of utopian and materialist attempts to refashion the social world, called for a political and economic transformation that would precipitate society into a life of harmony in the new industrial epoch.
> (Vidler 1992, 189)

The first of these utopias, the modern language of the art, has since the postmodern 'end of history' been subject to endless dissection, fragmentation and reassembly in a relentless search for meaning. The inheritors of Modernism's guiding instincts towards a positivist, scientific understanding of the city, latterly found themselves putting their faith in measurable traffic flows and spatial syntax. But now, within the Anthropocene, an appreciation of the complexity of the city, an ecosystem in its own right, gives renewed impetus to the second of these utopias.

> Le Corbusier's impatience at the obstinate survival of old Paris – 'Imagine all this junk, which till now has lain spread out over the soil like a dry crust, cleaned off and carried away' is only one extreme example of the radical shifts in sensibility toward the city in this (20th) century, shifts that have more often than not been resisted by the intractable nature of the existing urban fabric… A contemporary philosopher of urban architecture is faced then, at the end of the twentieth century, not so much with the absolute dialectic of ancient and modern posed by the avant- and rear gardes of the last eighty years, as with the more subtle and difficult task of calculating the limits of intervention according to the resistance of the city to change… In Hallbwach's words, "spatial images play such a role in the collective memory… a place receives the imprint of a group… a salutary caution to the urban architect in front of a seemingly passive plan of streets and houses".
> (Vidler 1992, 199)

The order and harmony of the modern, utopian world was to ensure that by clear expression of the elements of a programme, there would be an entirely appropriate visual order and harmony in architecture, free from ambiguity, since society itself would be by definition unified and unambiguous. Karl Popper strongly criticised the platonic values inherent in this kind of utopianism, which has been a recurrent theme throughout both revolutionary and totalitarian philosophies. The essence of utopian 'engineering' is that all its efforts are directed towards a single final goal – the ideal state. The first step is to clearly establish the form of this goal and then to formulate intermediate goals (e.g. the Soviet five-year plan) that must be reached to realise the final aim. In practice, there are problems that make the process impossible to carry through. First, since the ideal state lies in the future, its exact form is open to debate, any precise utopian vision is consequently limited to a single person or group of persons. Since they will find it difficult to convince everyone of the perfection of their proposals and since the intermediate steps are incapable of modification, they have to resort to oppression in order to push through their programme. Secondly, our lack of knowledge about the complexities of social engineering means that utopian visions are inevitably simplistic, and the familiar problems of implementation ensue. Unforeseen difficulties arise, the vision becomes compromised, and measures are put off until the utopian dream becomes the dystopian reality.

> This sweep, this extreme radicalism of the Platonic approach (and of the Marxian as well) is, I believe, connected with aestheticism i.e. with the desire to build a world which is not only a little better and more rational than ours, but which is free from all its ugliness; not a crazy quilt, an old garment badly patched, but an entirely new coat, a really beautiful new world ... this aesthetic enthusiasm becomes valuable only if it is bridled by reason, by a feeling of responsibility, and by a humanitarian urge to help. Otherwise, it is a dangerous enthusiasm, liable to develop into a form of neurosis or hysteria.
> (Popper 1945, 164–165)

Instead of utopian engineering Popper advocated 'piecemeal social engineering.' This rather than being concerned with the distant future, instead tackles immediate social problems. These can be more readily identified, and a consensus of opinion can be established as to how to deal with them. Inevitably, whatever we do will have some unforeseen results; in coming to terms with one problem, other problems will arise, so at each stage the situation has to be analysed and new goals established; it is then a continual process of experimentation and assessment. Similarly, within architecture, it suggests a continual process of feedback between designer and user, it is tentative and experimental rather than heroic and original, and realistic in the face of complex urban problems. Popper effectively describes the process of street-by-street retrofit, collaborative, step-by-step progress with no certain solutions or defined final outcomes; even as far back as 1970 *Meaning in Architecture* (Jencks and Baird 1970) noted the already threadbare justification for the first moderns' presumed gift of social prophecy.

The architecture of Functionalism replaced Classicism's references to order, hierarchy and inherited dominance, with faceless transparency. Bentham's paradigm of control, the Panopticon, had been revamped within the early twentieth-century public health agenda, such that it became representative of the new utopia based on scientific rationality. Jane Jacobs, notably, observed the social malaise resulting from the imposition of clarity, the destructive outcomes arising from attempts to 'clean-up' communities as a result of the articulation of function within cities. Amongst those who pushed back against Modernist doctrine, the Situationists constructing their mental maps of Paris sought to make a psycho-geographic image that captured the effect of locale on the behaviour of individuals, the mental life of the metropolis, capturing the psychological dimension of urbanity that modern functionalism denied.

But having eliminated the rich and decorated architectural language of classicism, elaborated over centuries, and replaced it with its converse, all that was available was a limited design palette. What once were cornices and mouldings, became shadow gaps, the recessed junctions of walls and doors. Colin Rowe and Venturi's solution was to once again engage with history, however: 'The ensuing attempt to rebuild the home on more stable foundations, according to the specifications of counter-Modernists and nostalgic dreamers, complete with its cellar and attic, its aged walls and comforting fireplace, has, however, inevitably fallen victim to a complaint inseparable from all nostalgic enterprises: that of the triumph of image over substance' (Vidler 1992, 66).

References

Graeber, D., and D. Wengrow. 2021. *The Dawn of Everything: A New History of Humanity*. London: Penguin Books.

Harari, Y.N. 2015. *Homo Deus: A Brief History of Tomorrow*. London: Harvill Secker.

Hinkel, J., D. Mangalagiu, A. Bisaro, and D. Tàbara. 2020. "Transformative Narratives for Climate action." *Climatic Change* 160: 495–506. Available online: https://doi.org/10.1007/s10584-020-02761-y.

Jencks, C., and G. Baird. 1970. *Meaning in Architecture*. London: Barrie and Jenkins.

Lent, J. 2017. *The Patterning Instinct: A Cultural History of Humanity's Search for Meaning*. New York, NY: Prometheus.

Nicolson, A. 2019. *The Making of Poetry: Coleridge, the Wordsworths and Their Year of Marvels*. Glasgow: William Collins.

Nicolson, A. 2020. 'In Wordsworth's Footsteps' BBC Sounds:3 *A Sense Sublime* 12 February.

Popper, K.R. 1945. *The Open Society and Its Enemies*, vol. 1. London: Routledge and Kegan Paul.

Stoknes, P.E. 2021. *Tomorrow's Economy, A Guide to Creating Healthy Green Growth*. Cambridge, MA: MIT Press.

Tafuri, M. 1976. *Architecture and Utopia: Design and Capitalist Development*. Cambridge, MA: MIT Press.

Vidler, A. 1992. *The Architectural Uncanny: Essays in the Modern Unhomely*. Cambridge, MA: MIT Press.

15 The Problem of Theory

Since World War II, architecture has split into many factions so that any unified theory has become ever more difficult, to the point now that architectural theory has all but ceased to exist. But at this point in time what is needed is a truly Green Architecture, not just a series of well-meaning, finger-wagging dictums, but a Green Retrofit Architecture that is a unified matter of belief.

The lack of a unified theory of architecture has repeatedly been discussed and lamented. The years following the 1960s saw a rash of manifestos promoting a common sense of direction. An ongoing commitment to modern architecture was accompanied by dissection of the shortcomings, social and urban, of the International Style. Inspiration has been drawn variously from semiotics, the pre-eminence of technology, and systems theory. But despite attempts such as Norberg Schulz's invocation of phenomenology, in *Genius Loci: Towards a Phenomenology of Architecture*, as an umbrella to contain architecture's disparate elements: 'After decades of abstract "scientific" theory, it is urgent that we return to a qualitative, phenomenological understanding of architecture' (Schulz 1979, 5). But after his book's long, nostalgic journey through architectural history, the Modernist orthodoxy, architecture's autonomy, is reasserted: 'This direction is not dictated by politics or science, but is existentially rooted in our everyday lifeworld' (Schulz 1979, 201). Since then, any hope of a combined theory has ceased to be in our 'post-critical' age governed by 'radical pragmatism.'

Eisenmann, invoking form as the basic criterion of architecture, has developed a language of grids, planes and columns, engaging with contemporary debate, but where those elements are shorn of any rational intent. Koolhaas ironically subverts the programme to achieve arresting contorted forms, an approach now shared with any number of contemporaries. Through the enjoyment of a larger palette of materials, Herzog de Meuron has achieved a further recycling of the Modernist project. The overriding aim, to reinvigorate Modernism, has to some extent been achieved, growing engineering sophistication providing the means for eye-catching, spectacular and often gravity-defying buildings. In all cases, the aestheticism of pristine form persists, the ravages of time and circumstance are ignored, yes, the Modernist slowing pulse has long continued but now, finally, seems without impetus:

DOI: 10.4324/9781003564997-19

'Contemporary architecture, in the most general sense of an art after Modernism, seems increasingly to be caught in the dilemma of Alice playing croquet with the Queen: she knew what the game was called but there did not seem to be any fixed rules' (Vidler 1992, 101).

Although Popper condemned utopianism per se, he didn't preclude the piecemeal social engineer from having visions of an ideal future. I.C. Jarvie (Jarvie 1987) expanded on Popper to suggest that piecemeal short-term planning could be supplemented by 'Critical Utopianism' – a range of possible futures rigorously criticised, and open to public choice so as to provide direction towards a long-term future. The goals established at any particular time would, however, have to be re-examined and redirected in the light of unforeseen events. This form of utopianism doesn't set down rigid goals but rather, general directions, which are of utility at that particular time. As he puts it: 'There is no reason why piecemeal planning should not follow on and benefit from utopian dreaming.'

Since then, scenario planning has become widespread from business to the Pentagon, a skill necessary and underlying successful retrofit, one to which the architectural mind is ideally suited, but in so doing casting aside our heroic persona for a caring alternative: 'Every innovation suffers by definition from a mismatch between the ways people currently do things and the ways they might do them. Open-ended in time means the way an object will evolve, how its use will change; the process cannot often be predicted in advance' (Sennett 2018, 14).

Utopianism has had a resurgence of interest in recent years, for while it may seem the stuff of naive myth, utopia is no more a myth than the one we currently live under, the myth of perpetual growth. That myth persists despite the all too visible private wealth amidst public squalor and the pervading climate unease. Yet a world without growth may soon be visited upon us; rates of growth across the Western world have inexorably been falling year on year. One theory maintains that the growth rates post-war into the 1960s may have been a one-off, the result of exploiting historically cheap fossil fuels. It has become all too clear that the blunt measurement of Gross Domestic Product that has been the economic standard for the last 50 years is a poor indicator of human health and well-being. Jason Hickel's *Less is More* (Hickel 2020) forensically dissects the future models that have been based on growth (including 'green growth') and all are found wanting, the only viable alternative is 'de-growth.' His title is of course well chosen, when Mies van der Rohe coined the dictum 'Less is More' it came to define the brave, utopian ideals of modern architecture. The massive difficulties in achieving 'de-growth' as outlined by Per Espen Stoknes: (Stoknes 2021) unemployment, the overhang of debt, the end of progress and innovation, are countered by Hickel in utopian fashion. He advocates: Ending planned obsolescence, cutting advertising, the sharing society, ending food waste, an end to ecologically destructive industry, reduced working hours and inequality, public provision of basic needs, and perhaps most challengingly the resetting of democracy. Sarah Ichioka and Michael Pawlyn in *Flourish* (Ichioka and Pawlyn 2021) adopt the same agenda with the addition of bio-mimicry and the circular economy, for the benefit of the built environment.

> Our love of nature: our innermost emotions; our empathy for others; our sorrow at the transience of life; our unfathomable connection to the earth itself; our capacity for contemplation; our inherited desire to strive for a better society. These are the psychological – and perhaps spiritual – dimensions of the human condition.
>
> (Jackson 2021, 66)

A common factor from Fourier onwards is the notion that in a utopian world, a balance should be found between what Hannah Arendt in *The Human Condition* (Arendt 1958) termed work and labour; labour being that which is a necessity for the maintenance of life – food, drink, shelter – including housework, parenting, care for the elderly. The reward of labour 'is the human way to experience the sheer bliss of being alive which we share with all living creatures.' Whereas work is about creation and creativity, it is what we do in order to attempt immortality by making our individual, lasting imprint; art is work. Rather than capitalist exploitation of labour – the excess value of labour, over and above that paid in wages, is given to employers and shareholders – in a utopian world, labour is divided equitably and equally within a community of sharing. Work is devalued by capitalism because the intention of work, to create permanence, is contrary to capitalism's need for constant change and novelty. Human needs aren't being met by capitalism, which exploits our instinctive drive to safeguard the future by acquiring and stockpiling an excess of goods and possessions. An economic system based on stress and relentless competition. A utopian sharing society would provide stability within which individuals would be free to achieve their full potential.

It is, as Will Self puts it: 'Civilisation has made us all house trained.' If we could be restored to our pre-industrial, indeed pre-agrarian lives, since when the power of rationality has become valued above the senses, the sensual and the sensuous, human creativity would be liberated. Obviously, this has implications for the world of architects who require, as Modernists, design to proceed logically from rational analysis of the project's programme. The alternative would be one valuing intuition and empathy and care. By rejecting the constraint of cartesian Modernism, a more intuitive feminine architecture concerned with the world as found, might be the outcome.

Richard Sennett in *Building and Dwelling* (2018, 9–10) unites the themes of Popper's Open Society, Hannah Arendt and the ideal represented by Homo faber, the craftsman, with the local activism of his friend Jane Jacobs, to write what could be a manual for the Retrofit Architect:

The role of the planner and architect would be to encourage complexity and to create an interactive, synergetic ville greater than the sum of its parts, within which pockets of order would orient people. Ethically an open city would of course tolerate differences and promote equality, but would more specifically free people from the straitjacket of the fixed and the familiar, creating a terrain in which they could experiment and expand their experience....

Popper abhorred economic monopolies just as he feared totalitarian states. Both make the same seductive promise: life can be made simpler, clearer, more user friendly as we would now say about technology, for example, if only people would submit to a regime which does the organizing. You will know what you are about, because the rules of your experience will be laid out for you. What you gain in clarity, however, you will lose in freedom...

To understand Homo faber's role in the city, we have to conceive of the dignity of labour differently. Rather than espousing a world-view, Homo faber in the city acquires honour by practising in a way whose terms are modest; the small house renovation done as cheaply as possible, or planting a street with young trees.

This is a model of practice adopted by Civic Square and Dark Matters Laboratory in Birmingham.

The politics of architecture

This (current) apolitical attitude is self-imposed, and stands in clear contrast to both the actual and the implicit claims of radical architects in the first decades of this (20th) century for urgent and active interventions, and to the more recent conciliatory calls for affirmation of the social and physical contexts as found. One suspects it is a symptom of the general and profoundly felt loss of the once-dominant tradition we now call the modern movement, of the prevalent cynicism about architecture's reforming and communicative powers, and of the general confusion about what it is that critically conscious architects, at this moment in history, should be doing.

(Hays 1995, 286–287)

In the twenty-first century, nearly all architecture has become commercial. Since the fall of communism, virtually the whole world has become governed by the dictates of neoliberalism. The market provides the programme for architecture, which in its relentless drive for profit eliminates the art within architecture or any other social good, through value engineering, and the lowest common denominator of design and build.

The insight that our way of life is destroying the ecosystem on which we depend has reanimated questions regarding the viability of 'capitalism' even in mainstream political debates ... A telling example is Naomi Klein's influential book: This Changes Everything, Capitalism vs. the Climate. Klein rightly emphasizes that the problem of climate change cannot primarily be addressed on an individual level but must be understood in terms of the economic system under which we live ... The large-scale questions raised by the problem of climate change should therefore be understood as a 'civilisational wake-up call. A powerful message – spoken in the language of fires, floods, droughts, and extinctions – telling us that we need an entirely new economic model and a new way of sharing this planet'. Rising to the challenges posed by climate change will require 'reinventing the very idea of the collective, the communal, the commons, the civil, and the civic' ... the capitalist measure of value is inimical to the production of real social wealth, since it valorizes socially necessary labour time rather than socially available free time, requires unemployment as a structural feature, and has an inherent tendency toward destructive crises.

(Hägglund 2019, 381–382)

Capitalism and creative destruction

It is easier to imagine the end of the world than the end of capitalism.

Fredric Jameson.

A familiar case made for capitalism is that competition drives innovation, accelerated by the ongoing exponential rate of technological change expressed within Moore's Law, so the destruction left in its wake is regarded as a force for good, as espoused by the economist Joseph Schumpeter (Schumpeter 2009), who having foreseen globalisation nonetheless also foresaw the end of capitalism. The bulging skyline of London, or any other financial centre, its towers torn down at increasing frequency to make way for yet taller monoliths, is one clear outline of creative destruction's power. If automation transforms both manufacturing and white-collar occupations, will this result in yet greater wealth inequality, as corporations grow larger still and finally surpass the power of governments, whilst caring occupations resistant to technological change, such as nursing, are further diminished? Alternatively, we await a revolutionary transformation, a utopian de-growth cultural shift that will inevitably undermine the current business model of the architectural profession.

> The pathos of all bourgeois monuments is that their material strength and solidity actually count for nothing and carry no weight at all, that they are blown away like frail reeds by the very forces of capitalist development that they celebrate. Even the most beautiful and impressive bourgeois buildings and public works are disposable, capitalized for fast depreciation and planned to be obsolete, closer in their social function to tents and encampments than to 'Egyptian pyramids, Roman aqueducts, Gothic cathedrals'.
>
> (Berman 1982, 99)

Modernism and radical politics

As an answer to the dilemma – should Retrofit Architecture be allied with ecomodernism – requires addressing the question of Modernism's roots in radical progressive politics. Giedion's version of the emergence of Modernism has technological progress as its driving logic, the same well that ecomodernism has as its source. But the Left can rightly lay claim to the Modernist project, at least in Europe, from the Bauhaus as a point of resistance to National Socialism, to the final emergence of Modernism in the United Kingdom, where it was allied to the post-war introduction of the welfare state. Subsequently, when Modernism's main proponents moved to the United States, Hitchcock and Johnson wrote 'The International Style' cutting ties with European utopianism. The United States became the centre of architectural development, Modernism became an aesthetic dominated by an avant-garde owing its allegiance to abstract, and later conceptual art, whilst the everyday practice of architecture retained its loyalty to orthodox Modernism. It is an irony that modern architecture has always sought to distance itself from the Arts and Crafts which is portrayed as the backward-looking originator of the hated suburbs. Yet both have left-wing beliefs as their underlying philosophy, can a future architecture continue to pretend that it is distinct from the world of politics?

References

Arendt, H. 1958. *The Human Condition.* Chicago, IL: University of Chicago Press.
Berman, M. 1982. *All That Is Solid Melts into Air: The Experience of Modernity.* New York, NY: Verso.
Hägglund, M. 2019. *This Life. Why Mortality Makes Us Free.* London: Profile Books.
Hays, K.M. 1995. *Modernism and the Posthumanist Subject: The Architecture of Hannes Meyer and Ludwig Hilberseimer.* Cambridge, MA: MIT Press.
Hickel, J. 2020. *Less Is More: How Degrowth Will Save the World.* London: Random House.
Ichioka, S., and M. Pawlyn. 2021. *Flourish.* Axminster, UK: Triarchy Press.
Jackson, T. 2021. *Post Growth, Life after Capitalism.* Cambridge, UK: Polity Press.
Jarvie, I.C. 1987. "Utopia and the Architect." In *Rationality: The Critical View*, edited by Joseph Agassi and I.C. Jarvie. Dordrecht: Kluwer Academic Publishers.
Schulz, C.N. 1979. *Genius Loci, Towards a Phenomenology of Architecture.* London: Rizzoli.
Schumpeter, J. 2009. *Can Capitalism Survive?: Creative Destruction and the Future of the Global Economy.* New York, NY: Harper Perennial Modern Thought.
Sennett, R. 2018. *Building and Dwelling: Ethics for the City.* London: Allen Lane.
Stoknes, P.E. 2021. *Tomorrow's Economy, A Guide to Creating Healthy Green Growth.* Cambridge, MA: MIT Press.
Vidler, A. 1992. *The Architectural Uncanny: Essays in the Modern Unhomely.* Cambridge, MA: MIT Press.

16 Architecture's Very Uniquely Compromised Position

> Capitalist realism takes the vanquishing of Modernism for granted: Modernism is now something that can periodically return, but only as a frozen aesthetic style, never as an ideal for living.
>
> (Fisher 2012, 8)

Compared with the other visual arts, architecture occupies a highly nuanced position, a delicate negotiation between the competing demands imposed by the planning system, the property market and the needs of the user. Many current commentators, such as de Graaf, have highlighted the distance between Modernism's original mandate and the hollowed-out language of pseudo-Modernism, the distant whisper remaining from the original clarion call. The transformational capability of the artist, espoused by Herbert Marcuse in *The Aesthetic Dimension* (Marcuse 1978), has rarely been possible in the world of built architecture, which has always been the servant of power; the paper architecture that emerged in the 1960s enabled the avant-garde to maintain its distance.

> The 20th century taught us that utopian thinking can have precarious consequences, but, if the course of history is dialectic, what follows? Does the 21st century mark the absence of utopias? And if so, what are the dangers of that? Piketty's framing of the 20th century echoes the familiar notion of 'the short 20th century': the historic period marked by a global contest between two competing ideologies, running from the beginning of the First World War to the end of Communism in Eastern Europe; starting in Sarajevo, ending in Berlin. If we are to believe Piketty, we may well be on the way back to a patrimonial form of capitalism. With that, modern architecture's social mission - the effort to establish a decent standard of living for all - seems a thing of the past. Architecture is now a tool of capital, complicit in a purpose antithetical to its erstwhile ideological endeavour.
>
> (de Graaf 2015)

That mission was clearly represented in the work of the politically committed from the first generation of Modernists, the architects of the New Objectivity. It might recur again in the retrofitted future city, where the modest improvisation of retrofit can present a vision of hope, the antithesis to the city of unregulated capitalism.

The urban landscape will then no longer serve as a site for intervention but become a cause for nurture and care. As with natural landscapes, cityscapes have their own moods, resonances and repositories of memory, the stand of trees that give their name to a crossroads, the street corner pub with a name derived from long-gone history. These markers were insignificant for post-war urban renewal, contemporary gated communities, or the gathering spaces in commercial redevelopments overlooked by CCTV and private security guards. Street-by-street retrofit for that reason needs close engagement with local people for whom their memories and markers may contain the essence of a place. Similarly, to the discovery of nature by the Romantics, this time providing psychological understanding of contemporary urban ecology, which isn't without its dark undertones. The contemporary psychogeography of Iain Sinclair similarly engages with the 'uncanny,' unescapable and unsettling aspects of the modern city, which are, as so often in literature and film, associated with malaise, epidemic, phobia and alienation.

> In the traditional city, antique, medieval, or Renaissance, urban memory was easy enough to define: it was that image of the city that enabled the citizen to identify with its past and present as a political, cultural and social entity; it was neither the 'reality' of the city nor a purely imaginary 'utopia' but rather the complex mental map of significance by which the city might be recognized as 'home', as something not foreign, and as constituting a (more or less) moral and protected environment for actual daily life.
>
> (Vidler 1992, 177)

The Modernist obsession with order and perfection, the notion that a work of architecture is a total and complete conception from which nothing can be added or taken away, in the platonic tradition linking Alberti and Classicism through to the modern, has left in its wake unresponsive and inflexible places that can't accept addition or change and has led to incoherent cities. An alternative philosophical tradition from Kant to Heidegger embodied the notion of the sublime (describing, for instance, the recent transformation of the City of London into a terrifying place of looming towers), and the 'uncanny,' the many disorienting locales that as in Freud's own dream, related in his '*The Uncanny*' (Freud 1919), seem impossible to escape. In the modern city 'uncanny' spaces abound, the delivery yards, back alleys and deserted fenced-in scrub land, what Ernst Bloch designated the 'hollow spaces of capitalism.'

> The contemporary sensibility that sees the uncanny erupt in empty parking lots around abandoned or run-down shopping malls, in the screened trompe l'oeil of simulated space, in, that is, the wasted margins and surface appearances of post-industrial culture ... aesthetically an outgrowth of the Burkean

Figure 16.1 One of Ernst Bloch's 'hollow spaces of capitalism'.
Source: Credit: the author

> sublime, a domesticated version of absolute terror … the uncanny as Walter Benjamin noted, was also born out of the rise of the great cities, their disturbingly heterogeneous crowds and newly scaled spaces.
>
> (Vidler 1992, 4)

The roots of Modernism: Hannes Meyer and the New Objectivity (Neue Sachlichkeit)

Throughout the twentieth century, there are many examples of politically committed architecture, both of the left and right. A precursor within Modernism of those that followed, such as the anti-architecture of the 1960s, was Neue Sachlichkeit, an example of what Charles Jencks termed 'The Activist Tradition' (Jencks 1973, 81). The New Objectivity emerged from the fervent debates of the 1920s about the role of the architect in the evolution of a new society, and the extent to which architects were to divine its essence and then represent it. The movement aimed for a universal objective, the realignment of the mental attitude of the times both in thought and feeling (Frampton 1980, 132). Nonetheless, its adherents derided the utopianism of Corbusier (Curtis 1982, 263–264), preferring to describe themselves as realists.

Underlying the technological and anti-art stance of the movement was the belief in a socially relevant architecture constructed to scientific principles, within which the individual was subject to the common good. As Hannes Meyer put it: 'Cooperation rules the world. The community rules the individual.' The Bauhaus was one of the battlegrounds between opposing architectural futures.

> While Le Corbusier maintained a distinction between the practical-technical role of the engineer and the artistic-poetic role of the architect in order to preserve the humanist autonomy of the latter, Meyer sought to eliminate traditionally conceived art altogether in favour of pure technique and the technical organization of a building in a collaborative enterprise. Thus, to Le Corbusier's 'engineering on the one hand, architecture on the other', Meyer would reply 'The new building is a prefabricated unit for site assembly and, as such, an industrial product and a work of specialists, economists, statisticians, hygienists, climatologists, industrial engineers, standards experts, heat engineers ... and the architect? He was an artist and has become a specialist in organization'. His reversal of the hierarchy of 'art' and techniques of mass production, that has routinely offended his critics ....
> (Hays 1995, 99)

Meyer the Marxist polemicist: 'architecture is not an aesthetic stimulus but a keen-edged weapon in the class struggle,' attracted vitriol, most notably from Gropius and Mies van der Rohe, but although dismissed as 'utilitarian' his architecture was one of great sophistication. His scheme for the League of Nations competition a worth contender with that of Le Corbusier.

Peter Bürger (Bürger 1992) explains that the intention of the avant-garde, and Hannes Meyer, was to reinvent the basis of art as understood through the lens of nineteenth-century aestheticism by denying its autonomy, thereby ridding high art of its 'highness,' its separation from everyday life and, in the case of architecture, establishing its engagement within the social sphere.

Gropius had followed an assiduously apolitical policy, whereas Oskar Schlemmer called the Bauhaus the: 'Rallying point for all those who, with faith in the future and willingness to storm the heavens, wish to build the cathedral of socialism.' Meyer was appointed head when Gropius resigned in 1928; he was sacked for his politics two years later and left for Moscow. Meyer's view has proved predictive, the architect is now but one of many specialists engaged in building projects, one searching for a lost role. The Retrofit Architect needs to be 'a specialist in organisation,' no longer just an artist, but socially engaged, an agent towards realisation of the new utopia.

City-scale retrofit offers, for architecture, the art most easily aligned to the needs of the market, an alternative mode of operation. Its canvas is the existing world and its amelioration. Rather than being that of the grand formal gesture, it offers up a new vision beyond the Modernist city, where the application of care, community choice and conversation generate a richer, sensuous vision, an alternative form of utopia.

As Herbert Marcuse wrote in *Art in the One-Dimensional Society* (Marcuse 1967, 119): '(The) image of art as technique in building or guiding the building of

148 *Street-by-Street Retrofit*

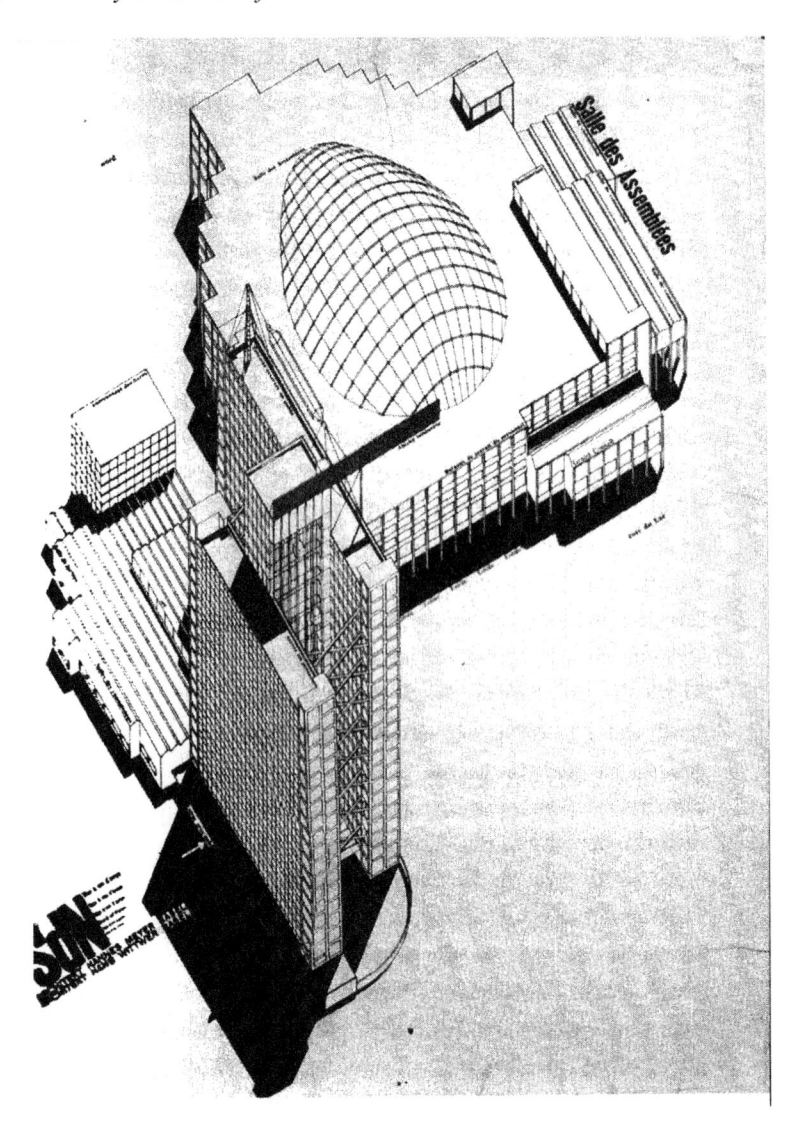

Figure 16.2 Hannes Meyer: Entry to the League of Nations Competition, 1926–1927.

Source: Credit: Reproduced from Claude Schnaidt, Hannes Meyer: Buildings, Projects and Writings (Teufen: Verlag Arthur Niggli, 1965)

the society calls for the interplay of science, technique and imagination to construct and sustain a new system of life. Technique as art, as construction of the beautiful, not as beautiful objects or places but as the form of a totality of life – society and nature.'

This isn't a clean sweep; we have to investigate history to find useful elements for the future.

The hidden aspects of consciousness, the uncanny, Gothic and Surrealism

> It is up to the left to adopt irrationality as a motor for the irrational act of compassion – the one political calling that can justify our empty human existence... To achieve this, we need an incendiary subcultural movement that rips up the political and cultural canons that led us to the impasse we're at ... Then we need to look outwards onto the streets in the formation of a grassroots movement that can reconfigure the debris of capitalism in new community-based formations.
>
> (Watson 2021, 101–102)

Despite the Surrealists being very individual characters, André Breton, their leader, travelled the world joining recruits to the movement, but he also expelled many along the way because he, with his demands for group solidarity, was building a political collective. The aim was to bring together Freud and Marx, to initiate human mental freedom concurrently with social revolution.

> 'Psychic automatism in its pure state, by which one proposes to express—verbally, by means of the written word, or in any other manner—the actual functioning of thought. Dictated by thought, in the absence of any control exercised by reason, exempt from any aesthetic or moral concern'.
>
> André Breton: Surrealist Manifesto (Breton 1945)

The Surrealists revelled in blurring the boundaries between the world of dreams and that of reality, between the organic and inorganic, opposing the technological rationality of Modernism and presaging current attempts to reintegrate humans and nature. Within modern architecture, of all the isms within twentieth-century arts, the psychic, dream world of Surrealism had limited impact, until its brief influence on deconstruction. Modernism resulted from within the scientific, positivist, tradition that emerged at the Enlightenment. This rationalist inheritance, which ironically it shares with the Chicago school of neo-liberalism, has obscured the non-rational sphere of existence, which fascinated the Surrealists.

As at the Renaissance, the example of Classicism with its orders and rules offered a precedent for a new highly structured architecture. As Colin Rowe demonstrated in *The Mathematics of the Ideal Villa* (Rowe 1976) this debt is often clearly expressed. For the Surrealists though, with roots within Dada and the absurd, any requirement for rules was an anathema. Instead, Surrealist inspirations are alchemy, myth and magic, the Gothic sphere of the imagination, which also resonates within the Arts and Crafts.

Discussing aesthetics, André Breton once noted that from a Surrealist point of view, the way in which a picture was painted was virtually irrelevant. It was the mental reality that the picture 'looked onto' that was all important.

The best way of appreciating the distinctiveness of Breton's attitude is by reference to the Modernist aesthetics of the critic Clement Greenberg, which dominated art criticism after Surrealism's demise in the late 1940s. Greenberg stressed disciplinary 'purity'. Far from opening onto some other reality, painting was required to draw attention to itself as an artistic discipline, to deal with problems intrinsic to painting. The rigours of this pursuit, Greenberg asserted, would lead to the attainment of degrees of formal or aesthetic, resolution.

(Hopkins 2004, 62)

This is still the dominant motivation within architectural practice. Each project has to achieve formal resolution with reductive consistency; each façade of the Modernist object has a limited vocabulary of repeated elements. These intentions accord perfectly with the demands of speculative finance and value engineering.

Situationism which by the mid-1960s had actually shed its artistic interest, could be understood as the inheritor of Surrealism's politics of everyday life. Opposed to the Modernist rationalisation of the city, the Situationists indulged in a kind of wandering, theorized by them in terms of 'dérive' or 'drift', and called in their writings for a 'unitary urbanism' whereby the spaces of the modern metropolis would be reorganized to fulfil the requirements of fantasy or play. There was a direct debt to Surrealism here. In an essay of 1950 entitled 'Pont-Neuf' André Breton had written of the ambiences attaching to streets in cities, such that in familiar streets one could demarcate 'zones of wellbeing and malaise'.

(Hopkins 2004, 151)

Giorgio de Chirico's metaphysical cityscapes, guided by Nietzsche's assertion that 'irrationalism (is) at the heart of Western culture' (Merjian 2014, 25), formed a pivotal influence on the Surrealist perception of the city, providing 'uncanny', distorted images that evade logical understanding. Whereas Modernism's 'logic is beauty' had led to the understanding of city as a system that can be reduced to equations, the Surrealists revelled in scenes that, happening by chance, evoked an emotive reaction.

For the Surrealists Paris was a city of revelation. Certain places acquired the character of avant-garde shrines ... As Surrealists they sought out particularly quirky or atmospheric sites ... Such locations seemed to be part of an alternative city; one hidden to tourists and miraculously spared by city planners, ruled by the logic of unconscious desire rather than day-to-day utility ... The Surrealists systematically allowed the city to penetrate their psyches; indeed, to map out their actions. By the early 1930s this was part of a self-consciously dialectical process ... Everyday events that appeared merely to be the outcome of accident – such as apparently chance meetings on the street – were deemed to correspond to psychic necessity.

(Hopkins 2004, 57–60)

Figure 16.3 André Breton in front of the painting by Giorgio de Chirico, "L'énigme d'une journée", 1922.

Source: © MAN RAY 2015 TRUST/ADAGP-ARS – 2024, image: Telimage, Paris. Copyright: For Man Ray – **© Man Ray 2015 Trust/DACS, London 2024** For de Chirico – **© DACS 2024**

Urban ecology

André Breton clashed with functionalist Modernism, for him 'psychic automatism' was to bring the creative process in line with explication of the unconscious, a psychic dimension lacking in CIAM's city of functional zones.

> Gradually generalized as a condition of modern anxiety, an alienation linked to its individual and poetic origins in romanticism, the uncanny finally became public in the metropolis … from the 1870s on, the metropolitan uncanny was increasingly conflated with metropolitan illness … the uncanny here became identified with all the phobias associated with spatial fear …

the uncanny emerged in the late nineteenth century as a special case of the many modern diseases, from phobias to neuroses, variously described by psychoanalysts, psychologists and philosophers as a distancing from reality forced by reality.

(Vidler 1992, 6)

The urban ecology has a psychic life of its own, places with memory, known or unrecognised, places with a sense of 'uncanny' unease, the deserted back lots and alleyways so well exploited by filmmakers and Edward Hopper. Surrealism encourages the Retrofit Architect to engage with these disorderly aspects of the urban scene, a dream life that the city has of its own. This agenda will require engagement with those rootless places – the suburbs and borderlands – so long spurned by architects. Modernist rationalism has treated the city as a matter of traffic flows, zoning and density distribution, those aspects that are readily measured and turned into mathematical models. Albert Einstein derided the dismissal of intuition; in the contemporary world, many of the senses that guided our ancestors, the sense of smell, for example, are diminished. Similarly, overlooked are the recollection and importance given to dreams in hunter-gatherer societies; despite their anti-rational fascination, dreams are mostly regarded as transitory, chance events, despite Freud and psychotherapy.

In the context of the nineteenth century city, the alienation of the individual expressed by writers from Rousseau to Baudelaire was gradually reinforced by the real economic and social estrangement experienced by the majority of its inhabitants ... urban estrangement was a consequence of the centralization of the state and the concentration of political and cultural power, where all local customs. and community bonds were brutally severed ... for Marx ... individual estrangement has become class alienation, the renter can never claim his home as his own.

(Vidler 1992, 4)

After the style wars of the nineteenth century, when for a period neo-Gothic was in the ascendancy, Modernism has often returned to Classical plan types, but Gothic with its air of spiritualism, mysticism and issues of faith and morality, had been consigned to history. Whereas the Greek and Roman orders are assumed to have been derived, highly abstracted, from earlier forms of wood construction, the Gothic cathedrals clearly have their roots in nature. Their branching dendriform columns reach ever higher to form a ribbed canopy beneath the sky. The master masons who gave their anonymous lives to the task had no concern for the passing of time, they weren't working to a Gantt chart, but to a higher ideal that guided by Gaia, will form the template for a future de-growth world working at a slower pace, what has been termed 'cathedral thinking'. As Gothic developed, over centuries, Romanesque giving way to Decorated and Perpendicular, naves grew taller and construction slenderer as masons learnt how to do more with less. Trial and error led to occasional collapse, but stones were reused, accumulating Ruskinian patina, and the other components made from trees and locally made glass, were found close by not with just-in-time supply chains crossing time zones. It is of

Figure 16.4 Carved capital in the form of hawthorn leaves, from the Chapter House at Southwell Minster, c. 1330.

Source: Credit: By Andrewrabbott – Own work, CC BY-SA 4.0, https://commons.wikimedia.org/w/index.php?curid=37918852

course the roots of Gothic in the other-worldly, the superstition supposedly overthrown by modern rationality and belief in technological progress, that has made Gothic an anathema. But recapturing its essence within nature may assuage the meaningless of modern human experience that lies at the root of contemporary discontents.

One episode, pre-dating the Romantics discovery of nature by several hundred years, is the story of the masons of York and Southwell Minsters who challenged the conventions of Early Gothic leaf carving, they went directly to nature for inspiration, the nature of the English countryside.

> And is not the balance of Southwell something deeper too than a balance of nature and style or of the imitative and the decorative? Is it not perhaps also a balance of God and World, the invisible and the visible? Could these leaves

of the English countryside, with all their freshness, move us so deeply if they were not carved in that spirit which filled the saints and poets and thinkers of the thirteenth century, the spirit of religious respect for the loveliness of created nature?

(Pevsner 1945, 66)

Pevsner's book relates the leaf carvings at Southwell to the number of encyclopaedic studies of plants, flowers and trees written at the same time, the time of St. Francis, that he suggests captures the spirit of the age, when 'cathedral time' included time enough for appreciation of the poetry of nature:

Be thou praised, O Lord, for our Sister, Mother Earth,
Who doth nourish us and ruleth over us,
And bring forth divers fruit and bright flowers and herbs.
From St. Francis's Hymn of the Sun.

The Arts and Crafts, although the only art movement with its origins exclusive to this country, has been subject to Modernist disdain, blamed for the vernacular of the suburbs, and arising as it did as successor to the Gothic Revival (Wilson 2007, 111). Lethaby's *Architecture, Mysticism and Myth* being an exemplar of the reviled superstitious past, to be swept away in the name of progress. Central to the Arts and Crafts was the pursuit of 'homeliness,' a retreat from the horrors of the nineteenth-century industrial city into the arcadian, safe, English countryside. Homeliness (*das Heimliche*) is the quality that Modernism finds hard to achieve, thus its unpopularity, whereas *das Unheimliche* (the unhomely) Freud's 'uncanny,' has thrived in the modern city.

... philosophers from Martin Heidegger to Gaston Bachelard wistfully meditated on the (lost) nature of 'dwelling', through nostalgic readings of the poets of the first, romantic uncanny. For Heidegger the unheimlich (unsettledness) was at least in his formulation of 1927, a question of the fundamental condition of anxiety in the world – the way in which the world was experienced as 'not a home' ... it was, of course, for ... security that, following the Second World War, Heidegger himself searched; attempting to trace the roots of pre-anxious dwelling and exhibiting a profound nostalgia for the premodern, his later writings have formed the basis of a veritable discourse on dwelling that has been taken up by the latter-day phenomenologists and post-Modernists alike.

(Vidler 1992, 8)

One of the founding fathers of the Arts and Crafts, Philip Webb, offers an insight into a now unfamiliar perspective on Architecture. His later buildings, after his first, the medievalist Red House (McNee 2012), are a highly eclectic exercise in modesty. As he put it: 'I cannot think that I consciously aimed at any particular

Figure 16.5 Four Gables, Brampton, Cumberland, c. 1878. Philip Webb's reinterpretation of a border country fortified tower house.

Source: Credit: the author

style; I by nature turned to romance rather than Classicalism and naturally without effort shrank from rhetoric.' His designs were rooted in the local vernacular whilst often, as in the case of his buildings in Cumberland (McEvoy 1989), achieving a startling grasp of the genius loci.

The 'awkwardness' of his architecture, a knowing and highly contrived artlessness, its studied ambiguity, is what distances it from its vernacular sources. Vernacular is considered the easy option nowadays, but Webb's vernacular was achieved only with the most rigorous attention to detail, he was happiest on-site immersed in the work of the craftsmen, truly a Master Builder.

> …though human beings need a sphere of independent action, and so of liberty, if they are to flourish, their deepest need is a home, a network of common practices and inherited traditions that confers on them the blessing of a settled identity. Indeed, without the undergirding support of a framework of common culture, the freedom of the individual so cherished by liberalism is of little value, and will not long survive.
>
> (Gray 1993, 309)

Standen, one of his last designs, although a large house, is planned along shifting axes that disguise its extent, maintaining the pretence of being linked enlarged cottages of different ages so as to achieve agelessness. The cottage was the scale at which he seems happiest to have worked, in order to achieve homeliness yet incompleteness; the junction with the existing farm buildings is effortlessly assured. Webb did not originate the idea of an architecture that pretended to have been added to and altered. It became a commonplace device for the later Arts and Crafts architects to construct a new building as dictated by its fictional history. For Webb, this was no mere picturesque whim but accorded with his intention of making a place-embedded architecture: 'I never begin to be satisfied until my work looks commonplace.' His aim was for his buildings to be rooted in their sites, even to the point of invisibility.

His reputation was achieved without his work ever appearing in magazines, and his motivations the exact reverse of the Modernist urge towards extravagant originality. Close colleague and fellow adherent of William Morris's socialism, co-participant in The Socialist League, unusually for the period, he designed staff quarters that were as carefully considered as those of his client. But the work of Webb and Morris rather than guided by the moral and spiritual dictates of the Ecclesiologists, was informed directly by nature. Since he shunned all honours, preferring membership of the more prosaic Sanitary Institute to that of the Royal Academy, Webb's practice was never lucrative. Whilst Webb and Morris extolled the craft tradition of the Middle Ages, they were at odds with the social and political organisation of that time, and increasingly with their own: 'Apart from the desire to produce beautiful things, the leading passion of my life has been the hatred of modern civilisation.'

Webb's lessons for the Retrofit Architect lie not in style but in approach, encapsulated in his commonplace work's modesty, invisibility, craft, seeming artlessness, agelessness, place-embedded, the very opposite of iconic.

> ...in Louis Kahn's favourite phrase 'What a building wants to be', the objective is not always something self-evident, particularly in times of change, it is probable that it will be concealed within a tangle of misunderstandings that requires patient elucidation and, like the Greek concept of truth, will have to be drawn out from concealment. Furthermore, when the state of affairs to be clarified is unprecedented, the search for techniques that will achieve the desired revelation will require a special kind of patience.
>
> (Wilson 2007, 87)

The paucity of methods and materials available to the retrofit designer has raised concerns about its impact on England's brick-built towns and cities. The concern is particularly that external render will, if implemented street-by-street, totally change the appearance of whole neighbourhoods. Or even worse, given the variety of tenure and ownership, rendered facades will randomly pepper-pot the street scene.

The means are both mundane and challenging. To make do and mend needs technology for reuse, current retrofit techniques, as observed earlier, are woefully lacking in terms of efficiency, environmental impact or fit within a circular economy, far from the processes of nature. Unlike the first moderns, for whom the technology was almost at hand, the new utopia labours with a lack of means – so inadequate compared with nature's reciprocal, cooperative and symbiotic world that utilises the power of the sun to create diversity, where all is recycled and nothing goes to waste.

References

Breton, A. 1945. *Primo Manifesto del Surrealismo*. Venice: Venezia Edizioni del Cavallino.
Bürger, P. 1992. *The Decline of Modernism*. Cambridge, UK: Polity Press.
Curtis, W.J.R. 1982. *Modern Architecture Since 1900*. London: Phaidon.
de Graaf, R. (2015), Architecture is now a tool of capital, complicit in a purpose antithetical to its social mission. The Architectural Review, 24 April.
Fisher, M. 2012. *Capitalist Realism: Is There No Alternative*. London: Zero Books.
Frampton, K. (1980), *Modern Architecture: A Critical History*, London: Thames and Hudson
Freud, S. 1919. *The Uncanny*. London: Penguin Books.
Gray, J. 1993. *An Agenda for Green Conservatism*. London: Routledge.
Hays, K.M. 1995. *Modernism and the Posthumanist Subject: The Architecture of Hannes Meyer and Ludwig Hilberseimer*. Cambridge, MA: MIT Press.
Hopkins, D. 2004. *Dada and Surrealism*. Oxford: OUP.
Jencks, C. (1973), *Modern Movements in Architecture*, New York, NY: Doubleday Anchor.
Marcuse, H. 1967. 'Art in the One-Dimensional Society', *Arts Magazine*, May 41:7.
Marcuse, H. 1978. *The Aesthetic Dimension*. Boston, MA: Beacon Press.
McEvoy, M. 1989. 'Webb at Brampton', Masters of Building Series: *Architects Journal*, 25 October.
McNee, P.E. 2012. 'The Red House is Not Perfect, but the Grace of Humanism is Visible in the Gleaming Oak Staircase', *Architects Journal*, November: 44.
Merjian, A.H. 2014. *Giorgio De Chirico and the Metaphysical City: Nietzsche, Modernism, Paris*. New Haven, CT: Yale University Press.
Pevsner, N. 1945. *The Leaves of Southwell*. London: Penguin.
Rowe, C. 1976. *The Mathematics of the Ideal Villa and Other Essays*. Cambridge, MA: MIT Press.
Vidler, A. 1992. *The Architectural Uncanny: Essays in the Modern Unhomely*. Cambridge, MA: MIT Press.
Watson, M. 2021. *The Memeing of Mark Fisher: How the Frankfurt School Foresaw Capitalist Realism and What to Do About It*. London: Zero Books.
Wilson, C. St J. 2007. *The Other Tradition of Modern Architecture: The Uncompleted Project*. London: Black Dog Publishing.

17 Echoes from the Past
Herbert Marcuse

One of Naomi Klein's key contentions is that if an ascent is to be made from the dystopia of the Anthropocene warming world, that the spirit of utopia will have to be rediscovered.

Herbert Marcuse, the Marxist philosopher of the Frankfurt School, achieved prominence in the 1960s as the students' guru during the failed revolution that sought to replace the existing order, on the campuses of the United States, and in 1968 on the streets of Paris. The other senior figures of the Frankfurt School, Adorno and Horkheimer, relocated back to Germany, but Marcuse stayed and

Figure 17.1 Herbert Marcuse addresses the Angela Davis Congress in Frankfurt, June 6, 1972.
Source: Credit: Keystone Press/Alamy stock photo

DOI: 10.4324/9781003564997-21

became deeply involved in student street protests alongside his protégé, the philosopher and prominent activist Angela Davis. Once again, we are at a juncture when many commentators believe change can't come from within the current system, only by an upsurge of public outcry will runaway capitalism be steered towards a post-growth future.

> It is Herbert Marcuse's utopia that has made him so significant and controversial a figure in our generation and possibly for generations to come.
>
> Marcuse endeavoured to describe a viable alternative to the present dilemma of civilisation through the projection of utopian ideals, and to liberate humans from the repressive forces of modern society in order to build a better world. Not surprisingly, his philosophical ideas took shape in relation to socialist and Marxist theory and its application to reality.
>
> One of the major aspects of this Modernist current seems to me to be The Spirit of Utopia, to use Ernst Bloch's expression. The present interest in utopia is no coincidence. Intoxicated by the frenetic pace of technology, and deprived of the traditional heritage of communal knowledge, contemporary man thinks he can see what Marcuse calls 'the end of utopia' – the possibility of realizing the wildest dreams of the utopians ... Utopianism ... defined the mental atmosphere of the New Left, of which, rightly or wrongly, Marcuse was considered a guru.
>
> (Martineau 1986, 3)

Marcuse reflected in his writing the optimism of the 1960s, he encouraged the belief that history does not inevitably lead to a given course, but that a new society could be realised, indeed that art could be a force for the negation of the status quo, and that society itself could become a work of art.

Society as a work of art

Marcuse conjectured on the possibility of 'society as a work of art,' as the aestheticisation of politics, that art (in the current context, the architecture of retrofit) was the means by which a future form of society could be imagined. It was a utopian ideal to end poverty and to issue a life of leisure that cut across class divisions. Art both acts as a critique of the existing social order, and he envisaged, through the imagining of an alternative, helps bring about its creation.

> For Marcuse, bourgeois culture generates popular forms such as the novel, but it uses them to offer a picture of life as it might be which is confined to the aesthetic realm. In this scenario, art is separated from the life-world in which its content might otherwise have agency. Bourgeois art displaces dreams of freedom to an immaterial realm where they do not threaten the existing social order.
>
> (Miles 2011, 52)

We are used to art as décor, art of rewarding but uncommitting value, pseudo-modern architecture. But for Marcuse, art can possess a 'convulsive beauty' with the power to shock, countering the narcotic comfort of bourgeois art and acting as the agency for social change. In bourgeois society, beauty, goodness and truth are subsumed within restricted cultural values that anaesthetise their original meaning; freedom is disguised as individualism, which is inherently unfree; capitalist economics and markets are the source of inequality. 'This ultimate principle of socialist theory is the sole absolute negation of the capitalist principle in all its forms .... Such freedom is the realisation of the fully developed needs, desires and potentialities ... liberation from the all-embracing apparatus of production, distribution and administration which today regiments ... life' (Marcuse 1998, 203).

Retrofit as the representation of society as a work of art

'Utopian representations knew an extraordinary revival in the 1960s; if post-Modernism is the substitute for the sixties and the compensation for their political failure, the question of utopia would seem to be a crucial test of what is left of our capacity to imagine change at all'.

(Jameson 1984, 16)

Marcuse asserted that this liberation was a necessity both from a human, sociological and political point of view and that if its achievement seemed utopian that was its strength because only in the complete negation of the status quo, the historic rupture, could there be 'a leap into the realm of freedom.' Given that the affluent American society of the time showed no sign of change, the working class having been captivated by the seeming satisfaction of material needs, the transformation would only take place once capitalism's contradictions became apparent. For Marcuse, the shift would happen by a complete reversal of the norm, the pursuit of freedom through work becoming play.

Marcuse attempted to say what a society as a work of art would be 'in concrete terms.' He speculates that play and imagination will reconfigure the cities and the countryside, restoring nature after its exploitation in capitalism ... the idea of a society in which work is play is not Marcuse's invention but an established concept in utopian thought since at least the early nineteenth century, where it figures prominently in the writing of Charles Fourier.

(Miles 2011, 101–102)

This is an important point for the realisation of a retrofitted urban realm. For this to be a successful project, it would have to be recognised as the rejection of current reality towards the achievement of a utopia. Retrofitting the domestic world will shed a light on the reality of the capitalist present, represented by its jostling shiny towers, repositories of speculative wealth. Only if the ecological, social and cultural goals of retrofit are adopted as a mass belief will the effort have been worthwhile in helping to restore the balance between the man-made and natural worlds.

For architects, the challenge of retrofit is to revitalise the moribund state of architecture, pseudo-Modernism, by radical imaginative engagement with the city as found. In the place of capitalist short-termism, speculative development for instant profit, new technologies will need to follow the regenerative processes of nature. The long life required of buildings in a de-growth world, and the uncertainties of climate change, will require ongoing adaptations. Buildings will need to slough off their skins and regenerate, a far cry from today's cycle of shoddy construction followed soon after by demolition and redevelopment.

Marcuse's utopia of hope, utopia as a realisable dream

'Hope is the pillar that holds up the world' Pliny the Elder

"This struggle to save the future," observes Al Gore, "will be played out in a contest between Earth Inc. and the Global Mind." The ease with which the internet transmits ideas across the world means that, when the time comes, the transformation of global consciousness could occur at a speed that might surprise everyone. It is part of our evolved human nature to stick together with our group's attitudes or opinions even when a changing situation leaves those attitudes out of date, which can frequently cause social rigidity and political inertia. When thought leaders emerge, offering new ways of thinking, they gradually attract increasing numbers of people until a tipping point and the old 'stickiness' that kept people attached to their old pattern of thinking is superseded by the pull of new ideas. All of a sudden, the gradual shift in ideas becomes an avalanche when those who are most comfortable sticking together find themselves in a rush to join in the new way of thinking. In the age of the internet, this tipping point can be reached much more quickly than in the past.
(Lent 2017, 439)

This describes a process similar to that of a 'paradigm shift' in scientific thought. Received wisdom, the accepted belief, can hold good until it is obvious that an existing theory is outmoded; the need for change becomes palpable. After a period of chaos, a new paradigm becomes an accepted belief. For Marcuse, this was the problem: how a new society can emerge from within the structures of the old, a utopia radically different from the past, yet limited by possibilities that can only be perceived within the terms of those that are currently available.

In Marx's view, it is vain to expect that that any important change can be achieved by the use of legal or political means; a political revolution can only lead to one set of rulers giving way to another set ... Only the evolution of the underlying essence, the economic reality, can produce any essential or real change – a social revolution. And only when such a social revolution has become a reality, only then can a political revolution be of any significance. But even in this case, the political revolution is only the outward expression of the essential or real change that has occurred before. In accordance with this theory, Marx asserts that every social revolution develops in the following way. The material conditions of production grow and mature until they begin to conflict with social and legal relations, outgrowing them like clothes until they burst ... it is I believe, impossible to identify the Russian Revolution with the social revolution prophesized by Marx, it has in fact no similarity with it whatsoever.

(Popper 1945a, 119–120)

Retrofit as subversive art

Marcuse's life-long insistence on the radical potential of art is linked to (his) obstinate insistence on the utopian dimension. On the one hand, art criticises and negates the existing social order by the power of its form, which in turn creates another universe, thus hinting at the possibility of building a new social order... On the other hand, emancipatory possibilities reside in the very forces that are responsible for the obscene expansion of an increasingly exploitative and repressive order.

(Davies 2004, 46)

Art's ability to stand outside the established mindset could entail the anti-art of Dada or the artifice of Surrealism. Marcuse argues that a sense of the political is inherent to art, that it is the aesthetic form of art that is autonomous, standing apart from the conditions of its production and thereby able to subvert society's prevailing consciousness. He maintains that it is in the realm of imagination that art is freed from the conformity of capitalist society and instead becomes the agency for production of a new dimension of thought, the first incidence of an emerging reality. Marcuse viewed art as presenting the radical alternative, the complete negation of a bureaucratic society of suffocating administration, a state only too familiar under neoliberalism. Art's imagination counters the constant assertion by those in power that 'there is no alternative.'

For the retrofit architect, the need to embrace an unknown future through time, requires a modesty of intentions, met through caring engagement with communities and individuals. Nonetheless, the task comes with the responsibility to persuade and guide residents towards alternatives of which they might not be aware. The role is however that much larger, requiring architects to be agents of the societal change that will come about by social revolution, that only through a sea change in public attitudes will utopia be achieved. The outcome, a patchwork retrofitted city of opportunity, reflecting diverse patterns of life, human and non-human, respecting unquantifiable qualities of place, will represent within it: 'society as a work of art.'

Art interrupts and is potentially convulsive. This is a romantic idea, but also millenarian, echoing 'the libertarian sects of the Middle Ages' and the ideas of Charles Fourier as well as early Marx. But is it still relevant today? Perhaps the realities of climate change, mass migration and wars over water in the mid to late twenty-first century will provoke a millenarian upheaval.

(Miles 2011, 146)

References

Davies, A.Y. 2004. "Marcuse's Legacies." In *Marcuse: A Critical Reader*, edited by J. Abromeit and W.M. Cobb, 46. New York, NY: Routledge.
Jameson, F. 1984. *PostModernism, or, The Cultural Logic of Late Capitalism*. Durham, NC: Duke University Press.
Lent, J. 2017. *The Patterning Instinct: A Cultural History of Humanity's Search for Meaning*. New York, NY: Prometheus.
Marcuse, H. 1998. "Technology, War and Fascism." In *Herbert Marcuse's Collected Papers*, vol. 1, edited by D. Kellner and P. Marcuse. London: Routledge.
Martineau, A. 1986. *Herbert Marcuse's Utopia*. Eugene, OR: Harvest House.
Miles, M. 2011. *Herbert Marcuse – An Aesthetic of Liberation*. London: Pluto.
Popper, K.R. 1945a. *The Open Society and Its Enemies*, vol. 2. London: Routledge and Kegan Paul.

18 Retrofit and Architects

A Future

> Design always has had to achieve multiple goals; the importance of climate change means that energy and CO_2 emissions have to have a high priority... As a result, the relationship between the architect and the building services engineer is changing: function has to precede form.
>
> (Boardman 2005)

The Green Deal from ten years ago was shunned by architects (Murray 2013) and was anyway a dismal failure. The concern then was how architects, most of whom operate as the smallest of SMEs, would avoid being excluded from retrofit, once it was rolled out at a larger scale and taken over by large corporations offering a one-stop-shop to households. A lack of sensitivity to local design considerations and the circumstances of individual houses and householders sank the Green Deal. It has since become apparent, given the number of technical failures, that market-fuelled retrofit tends towards the lowest common denominator, guaranteed by a lack of choice. Particularly through the Retrofit for the Future Competition, some architects committed to retrofit had a guiding role as the first port-of-call within communities. But if the 'national war effort' to retrofit the country ever comes to pass, we can expect under the current regime that big construction companies and the usual market forces will prevail.

The existing process of retrofit is governed by the availability of funds and maintenance cycles. A property's retrofit strategy will need to envisage a solution for new windows when wall installation is installed, an appropriate method of ventilation according to the air permeability of the building, and heating systems that match the response time and thermal capacity post-retrofit. Architects, well versed in small residential projects, are ideally placed to design these incremental retrofits and to specify and manage the integrated packages of improvements that will be required. Otherwise, in a low-margin market for retrofit, an uncoordinated approach is unlikely to achieve quality or satisfactory outcomes.

The current solution within PAS 2035, recognising the pitfalls and complexities of the task, is the need for skills beyond those of an advisor, those of Retrofit Coordinator, a skill set interchangeable with those of an architect. Project management abilities are required to overcome the potential pitfalls, particularly the inherent

DOI: 10.4324/9781003564997-22

dangers of condensation and poor indoor air quality (as investigated in Part 2). The Retrofit Coordinator is to provide liaison with occupants from start to finish and smooth the usually fragmented process of delivery. This sub-profession is to develop the still meagre basis of retrofit knowledge and encourage the long-term benefits of deep retrofit through post-occupancy evaluation and dissemination of best-practice. Would this extra string to the architect's bow help recover some of the credibility and capability that has been lost to the profession in recent times? Yet architectural enthusiasm remains dim, despite the enormous task that street-by-street retrofit will present. Retrofit for the architectural press still means historic brick warehouses converted into art galleries.

But the as yet unrecognised dimension is that in a de-growth world, one of make-do-and-mend, making do and making good over generational time spans, wholesale retrofit comes to the fore. This is not to say that the city scene won't change and mature as the social revolution, society as art, the utopian future itself matures, the city scene reflecting a communal, egalitarian, diverse, non-competitive and non-combative world governed by feminine values, care, empathy and intuition. How will this look and come about when the technical means are not to hand, available technologies are a long way from the processes of nature. The application of architectural imagination and ingenuity is urgently required to bring about the new utopia.

Architects and innovation – Our utopian mission

Here in 2024, the challenges are no different from ten years ago. For countrywide retrofit to happen, SMEs will be the key, but will architects just be mere sub-contractors? A better route would involve the Architect/Retrofit Coordinator providing a guiding role throughout the process to iron out the lack of responsibility taken at the junctions between trades, a principal reason for projects not performing to expectations. Householders need to be presented with integrated plans for the measures required, clear and simple information for their homes' future use, with quality assurance and risk management designed in. The alternative is the continuation of retrofit's recent history of thermal bridges, air leakage, condensation and heat loss. The Retrofit Architect would be best placed to set up post-occupancy evaluations and monitoring, a challenge in a high-volume, low-margin business but helping achieve the holy grail of grasping back architect's traditional expertise and giving increased and enhanced profile to the profession.

The need is for practical concerns with the logistics and costs of building to be paramount because retrofit architects, centred within communities, will have to 'get their hands dirty.' Most of the country's retrofits will have to be carried out while homes are occupied, with consequent issues of achieving construction on time and on budget. Helping households bear the upheaval while helping their retrofitted houses to work to plan will require the best of communication skills. For sure, not all whole-house upgrades will happen at the same time, despite the urgency of the task. Housing associations can build phased works into their asset management plans to carry out measures that naturally coincide, such as replacement windows with external wall insulation, and internal insulation with replacement kitchen units. Owner-occupied housing will need the progressive introduction of an improvement plan, a shopping list of low energy measures and priorities, for every home in the country, and a statutory document that would be required alongside its Energy Performance Certificate.

The role of Retrofit Coordinator is a clear fit with architectural skills through stages of assessment, design, planning submission, specification, tendering, contract administration and post-occupancy evaluation. A holistic role that encompasses cost advice, judging the balance of risk and benefit, and maintaining communications and control of quality. The advantages of employing this person are those that architects would naturally advocate for themselves – experience of practical solutions, communicating and defining responsibilities, risk management, reporting and planning the following stages through completion. The skills required will grow with scale, whole estates and streets will need particular concern for qualities of place to be sympathetically understood. The match with architectural skills is so exact that it is foolhardy for architects to stand aside from the process, but the implications for architectural education are profound: 'The profession, which includes the pedagogical frameworks of most schools of architecture and their ties to the professional accreditation system, needs to shift from defining itself solely through the act of designing buildings and instead engage with the much wider social, ecological and political systems that make up the production of the built environment' (Till and Schneider 2011).

What for society is a time of agonising change, represents for architecture an opportunity to become recombined with its social purpose: To be the built representation of a truly, open society. Architects embracing the spirit of a new idealism, a new utopian dream, become agents of change, the retrofitted city becoming the principal cultural manifestation of 'society as a work of art.'

> Because of our unique cognitive capacity, human social systems need to be understood as a pair of two tightly interconnected, coexisting complex systems: a tangible system and a cognitive system. The tangible system refers to everything that can be seen and touched: a society's tools, its physical infrastructure; and its agriculture, terrain, and climate, to name just some of its components. The cognitive system refers to what can't be touched but exists in the cognitive network of the society's culture: its language, myths, core metaphors, know-how, hierarchy of values, and worldview. The coupled systems interact dynamically, creating their own feedback loops, which can profoundly affect each other and, consequently, the direction of the society… There seems little doubt that we are currently in the midst of one of the great critical transitions of the human journey, and yet it is not at all clear where we will end up once our current system resolves into a newly stable state.
>
> (Lent 2017, 25)

Central to Modernism, the machine analogy, has been brought to bear on everything, including the organisation of human affairs and economies; it has dominated our understanding of nature itself and underlies the dispiriting aesthetic of pseudo-Modernism. The advances and advantages that have been realised by Modernism are now clearly bumping up against the constraints of the natural world. Science, so widely admired by the first modern architects, can be seen as presenting as many problems as solutions, now that a technological fix is being grasped as the only solution to the climate problem, with the gigantic attendant risks of geo-engineering. For humanity, to manage without the conveniences of modern technology would be unthinkable, but we can already see the emergence of the sharing society and new forms of virtual community. To bring the whole world population to the current expectations of the West's middle class will be beyond the capacity of the planet. But the retrofit social revolution, led by architects, offers to the world one beguiling aspect of a utopia that manages with less, but for quality of life, and the city, offers much more.

The architects of the 1920s created a future that was a utopian dream; we too need a utopian guide. The aim of these last chapters has been to find those themes and isms from the past that might help find this direction, give current realities: The relationship of pseudo-Modernism to power and politics, the means by which architecture can be wedded to the cause of a new utopia and a new economic order. 'Sustainable development' is clearly a nonsense – one of the 'hypnotic-ritual formulas of present-day neo-conservatism and neo-liberalism' (Marcuse 1978, 104). Our aim in a de-growth world has to be to manage equitably, with less not more, making an architecture of re-use for a sharing society wedded to both survival and human joy and pleasure. Modernism has made for a rootless world that does not achieve psychological security; the notion of progress concurs with a questionable, ecomodernist perspective; female modes of operation concerned with empathy, intuition and care are the way to a better future.

'A great epoch has begun. There exists a new spirit' Le Corbusier: Towards a New Architecture.

For architects, city-wide retrofit offers a view into a different belief, that the city is itself an ecosystem within nature. Architectural theory has exclusively been concerned with architecture as a visual art, but cities have qualities of remembrance that involve the totality of the senses; old cities are like suits of clothes that may be threadbare but comfortable and familiar. The retrofitted city won't ever be complete and not a 'work of art' as understood by Clement Greenberg, but rather Karl Popper's 'crazy quilt.' As the conception of time changes in a de-growth world, the maturing complexity of cities and their intricacy become qualities to admire. So, for architects, retrofit is an opportunity to reconnect; but imbued with the spirit of change, the new utopia, they become agents of change, a role far more profound than that purely instrumental task of retrofit coordinator.

It is not yet clear, however, whether any democratically elected leader will be able to marshal levels of popular support sufficient to uproot industries and ways of living that must be ended if global warming is to be stopped.
(Gerstle 2022, 285)

Ultimately, art needs to be opposed to the drudgery of life under capitalism, with the former's promise of freedom being writ large all over our global banks and national stock exchanges. We need a Marcusian Great Refusal for our times.

(Watson 2021, 100)

Architecture or extinction

References

Boardman, B. 2005. *Examining the Carbon Agenda via the 40 Per Cent House Scenario*. Oxford: Environmental Change Institute, University of Oxford.
Gerstle, G. 2022. *The Rise and Fall of the Neoliberal Order, America and the World in the Free Market Era*. New York, NY: Oxford University Press.
Lent, J. 2017. *The Patterning Instinct: A Cultural History of Humanity's Search for Meaning*. New York, NY: Prometheus.
Marcuse, H. 1978. *The Aesthetic Dimension*. Boston, MA: Beacon Press.
Murray, C. 2013. 'This Poorly Thought-Out Green Deal is a Bad Deal for Architects', *Architects Journal*, 31 January: 5.
Till, J., and T. Schneider. 2011. "Beyond Building." *Architecture Today* 218: 10.
Watson, M. 2021. *The Memeing of Mark Fisher: How the Frankfurt School Foresaw Capitalist Realism and What to Do About It*. London: Zero Books.

Index

Note: *Italicized* pages refer to figures.

Achieving Zero (Boardman) 69
'Active House' approach 50
'The Activist Tradition' (Jencks) 146
adherence loop 113–114
AECB Retrofit Standard (AECB 2021) 52
Aereco 82–83
aerogel 68
The Aesthetic Dimension (Marcuse) 144
The Age of Spectacle (Dyckhoff) 5
air-conditioning 70
air source heat pumps (ASHP) 98
airtightness 89–94
All Hands to the Pump, A Home Improvement Plan for England (IPPR 2020) 55
Anthropocene 26, 29, 131, 135, 158; and nature 12–13
Anthropocene Architecture School 133
architects: of 1920s 167; future of 164–168; future role of, as building sparingly 36–37; and innovation 165–168; as retrofit leaders 126
Architects Climate Action Network (ACAN) 31, 110
architectural skills 126–127
architecture: as a belief system 30; politics of 141–142; retrofit 37–38
An Architecture Degrowth Manifesto 133
Architecture, Mysticism and Myth (Lethaby) 154
architecture of Functionalism 136–137
Architecture's Evil Empire: The Triumph and Tragedy of Global Modernism (Glendinning) 5
Art in the One-Dimensional Society (Marcuse) 147–148
Arup 56, 88, 89

ASHRAE 90
Australia, solar panels 103

Bachelard, G. 154
Bankers without Borders 57
Bauhaus 1, 143, 147
Being Ecological (Morton) 29
Bere Architects' 46
Better Building Partnership (2012) 48
Birmingham Energy Savers (BES 2012) 96
Bloch, E. 145, 159
Boardman, B. 61–62, 69, 82, 124
Boiler Upgrade Scheme (2022) 124
Brand, S. 13, 22, 78, 91, 125
BREEAM Refurbishment and Fit-Out Framework 50
Breton, A. 150–151, *151*
Brundtland Commission 8
BSkyB's ventilation chimneys 46
Building and Dwelling (Sennett) 140
Building Automation and Control Systems (BACS) 83, 99
Building Integrated PV (BIPV) 100
Building Research Establishment Domestic Energy Model (BREDEM) 43, 115
The Building Regulations 44, 47–48, 51, 71, 78–79, 81, 83, 90, 93, 96, 105, 112
Building Safety Act 37
Bürger, P. 147
'The Business Case: Incorporating Adaptation Measures in Retrofits' (London Climate Change Partnership 2014) 56

Capital in the 21st Century (Picketty) 6
capitalism 6, 7, 28, 36, 55, 106, 140, 142, 145, 159; and creative destruction 142

Index

CarbonBuzz 89
CarbonCulture 113
carbon dioxide emissions 28, 97, 131
carbon neutral 53
caring architecture 35–38
carrying capacity of planet 12–18; Anthropocene and nature 12–13; climate crisis 14; cultural juncture 14–15; Ecomodernism and magic of technology 13–14; sustainable development origins and implications 13
carved capital in the form of hawthorn leaves *153*
case studies: Energiesprong 9–11; IFORE Innovation for Renewal 15–18; Link Road, Birmingham 31–33; Parity Projects 23–25
Centre of Refurbishment Excellence (CORE) 107–108
CERT 10, 97, 123
CESP 10, 123
circular economy 7, 35, 51, 54, 140
Civic Square 33, 141
Clean Heat Grant (April 2022) 56
climate change 8, 22, 31, 43, 54, 67, 71, 77, 111, 114, 120, 123–125
Climate Change Act (2008) 76, 86
Climate Change and the Indoor Environment: Impacts and Adaptation (CIBSE 2005) 71
Climate Change Committee 110, 111
climate crisis 14, 56
Climate Emergency Design Guide (LETI 2020) 51–52; proposals for retrofit (2021) 52–53
Climate Leviathan (Wainwright and Mann) 7
Coefficient of Performance 98
Commons Environmental Audit Committee 110
Community Empowerment Networks 76
Community Energy Savings Programme 76
Community Interest Companies (CICs) 110
cost-benefit analysis 56
Cost of Poor Housing Report (BRE 2023) 42
Critical Utopianism 139
cultural juncture 14–15

Dark Matters Labs (Dm) 31–33, 141
Davis, P. (Princedale Road) 87, 93
'decarbonisation' of electricity grid 127
Decent Homes Standard 42
decentralised constant mechanical extract (dCME) 90

de Chirico, G. 150
decision-making 38
de-growth 7, 132, 139, 142, 152, 161, 165, 167
de Graff, R. 6
'demand control' system 79, 80, 82
Department of Energy and Climate Change (DECC) 89
determining outcomes 44–57
Domesday Book (new) 37, 123
domestic heating carbon emissions 111
domestic refurbishment standard 50
Domestic Renewable Heat Incentive (RHI) 56
Dorling, D. 28–29
Dunster, B. 46
Dyckhoff, T. 5
Dynamic Insulation 65, 83

Early Gothic leaf carving *153*
ECO 10, 97
ECO+ 124
ecomodernism and magic of technology 13–14, 143, 167
Einstein, A. 36, 152
Eisenmann, P. 138
Electricity Act of 1947 101–102
Encraft engineering consultancy 73
Energy Efficiency: Recalibrating the Debate (EST) 109
Energy Performance Certificates (EPCs) 42, 50, 56–57, 77–78, 88, 96, 112
Energy Saving Bundles 87
Energy Saving Trust (EST 2010) 83, 98, 109
Energy Systems Modelling Environment (ETI) 110
EnerPHit 45, 47–49, 51–52, 63, 66, 84, 93
English House Conditions Survey (GOV.UK. 2006) 77
English Housing Survey 2020–21 42
EPC Action Plan 112
Eros and Civilisation (Marcuse) 30
ESCO 104
ethical capitalism 55
EU Interreg project IFORE 15–18, 107, 109
European Green Deal's Renovation Wave 111
external operable blinds 73
external wall insulation (EWI) 63–64

'fabric first' 49, 62, 70, 105, 116
Fanger, O. 71, 115
feed-in-tariff (FIT) 102–103

Flourish (Ichioka and Pawlyn) 139–140
The 40 per cent House (Boardman) 61–62, 69–70, 77–78
Four Gables, Brampton, Cumberland *155*
Fourier, C. 162, 140, 160
Four Walls and a Roof (de Graff) 6
fuel poverty 16, 43, 52, 55, 70, 76, 102, 107, 110, 115, 117
Future Energy Scenarios 111
FutureFit report (Affinity Sutton 2013) 116–117
Future Homes Standard (2025) 79, 80, 112

Gaia 35, 131, 152
Gaia hypothesis 26, 29
Genius Loci: Towards a Phenomenology of Architecture (Schulz) 138
Gentoo 87
geo-engineering 22
Getting People Involved (PlanLoCaL 2013) 119–120
Giedion, S. 22, 143
Glasgow Caledonian 44
Glass-Xcrystal 74
Global Footprint Network 8
global warming 12, 31–32, 42–43
Golden Rule (Green Deal) 87
Greenberg, C. 150, 167
'green capitalism', control of 8
Green Deal 78, 96, 164; 2009 87; 2012 127; failure of 97; 'pay-as-you-save' (PAYS) 109–110
Green Deal Finance Company 87
Green Deal Pilot scheme 87
Green Doctor 18
'green growth' 7
Green Homes Grant Local Authority Delivery Scheme 56
Green New Deal 103
Greta 104
'grey growth' 7, 8
Gropius, W. 134, 147
Gross Domestic Product (GDP) 7, 12, 30
Grosvenor Estates (2019) 43–44
Ground Source Heat Pumps (GSHP) 98

Hard to Decarbonise Social Homes (LGA 2022) 123
Heat and Building Strategy 56, 72
Heat and Energy Saving Strategy (GOV.UK. 2008) 76, 88
Heat Loss Parameter 112
Heat Markets Forum 76
Heidegger, M. 154

HEPI Solar project 104
Herzog de Meuron 138
Heschong, L. 84
Hitchcock, H-R. 1
hollow spaces of capitalism (Bloch) 145–146, *146*
Home Energy Master Plan (HEMP) 89
Homes Upgrade Grant 124
Home Truths (Boardman) 69, 76, 77
Homo Deus (Harari) 131
Hopper, E. 152
Human Condition (Arendt) 140
hybrid ventilation 84

Ichioka, S. 139
indoor air quality (IAQ) 89–90
Industrial Revolution 8
innovation, and architects 165–168
Institute of Sustainability 86–88, 92, 115
insulation 62–68; for cavity walls 64; internal wall 64–65; Parity Projects 2009 67; Rockwool 65; STO-solar 65; technology development in 67; use of robots in 65
intermittent ventilation 78, 79
internal blinds 73
internal wall insulation (IWI) 43, 64–65
The International Style (Johnson and Hitchcock) 1, 138, 143
Inventers system 78

Jackson, T. 28, 36
Jacobs, J. 19, 137, 140
James, H. 133
Jameson, F. 20, 142, 160
Jarvie, I.C. 139
Jencks, C. 146
Jevon's paradox 115
Johnson, P. 1

Klein, N. 141–142, 158
Koolhaas, R. 138

Lambeth Skyroom project 37
Last Child in the Woods: Saving our Children from Nature-Deficit Disorder (Louv) 29
Le Corbusier 19, 27, 93, 135, 146
LEO (Project Low Energy Oxfordshire) 104
Less is More (Hickel) 139
Lethaby, W. 154
Limits to Growth (Randers) 7
Link Road, Birmingham 31–33

lithium-ion batteries 101
Local Authority Revolving Fund 110
London Building Stock Model 37
London Climate Change Partnership (2013) 54
London Energy Transition Initiative (LETI) 10–11; methodology 53; whole building retrofit plan 53
London Homes Energy Efficiency Programme (LHEEP) 108
Loughborough University 72
Louv, R. 29
Lovelock, J. 8, 26
low carbon Britain 110–111
low-carbon technology 70
Low Carbon Transition Plan 2009 (GOV.UK) 69, 86
Lunos system 82
Lynas, M. 26

Machine for Living in (Corbusier) 93
Mackintosh Environmental Architecture Research Unit (MEARU) 91
Malm, A. 20, 36
Manhattan Project (New) 23, 26, 43, 68
Mann, G. 7–8
Marcuse, H. 30, 144, 147–148, *158*, 158–160; utopia of hope 161–162
Martin, L. 1
materials resource banks 51
The Mathematics of the Ideal Villa (Rowe) 149
Meaning in Architecture (Jencks and Baird) 136
Meyer, H. 146–148, *148*; Entry to the League of Nations Competition (1926–1927) 147, *148*
Microgeneration Certification Scheme (MSC) 101
MINDSPACE (IfG 2010) 114
'The Missing Quarter' project 113
'modern' architecture 135–136
Modernism 1, 138; geo-engineering 22; and natural world 29–30; and radical politics 143; re-evaluation of 19–25; roots of 146–148, *148*; self-regulation 20–21
Modern Methods of Construction (MMC) 10
Moneo, R. 5
monitoring 10–11, 24, 36, 49, 52–53, 61, 65, 86, 99, 107–108, 112, 165
Moore's Law 13, 142
Morris, W. 156

Morton, T. 22, 29, 38
MVHR (Mechanical Ventilation Heat Reclaim) 45–46, 48, 78–81, 87, 93

NASA's Jet Propulsion Laboratory 68
National Energy Efficiency Data-Framework (NEED) 37
National Energy Study (Sustainable Homes 2014) 116
National Grid 111
National House Building Council (NHBC) 46–47, 80, 82, 113
National Refurbishment Centre (2012) 89
National Retrofit Strategy 110
nature, Anthropocene and 12–13
neoliberalism 6, 33, 141, 149, 162
'net zero' building 51, 53
net zero carbon reduction 111
Net Zero Strategy: Build Back Greener (GOV.UK. 2021) 69
Neue Sachlichkeit (New Objectivity) 146
New Economics Foundation's Social Return on Investment tool 56
Nicolson, A. 133

occupants 111–120
Ofgem 103
overheating 70–74
over-ventilation 94

paradigm shift 161
Parity Projects 23–25, 88–89; 'Portfolio' Stock Assessment Tool 123
Passive House Institute 79–80
Passivhaus: 16, 22, 45–52, 66–67, 71–73, 79–82, 93, 98, 115, 124; overheating criterion within 71–73
Pawlyn, M. 139
Pay-As-You-Save (PAYS) pilot scheme 87, 110
pepper-potting 17, 156
performance gap 47–48, 112
petrochemicals 67–68
Pevsner, N. 153–154
phase-change materials (PCM) 72
photovoltaics (PV) 70, 96, 99, 100
PHPP 45–49, 52, 72, 107
Picketty, T. 6
piecemeal social engineering 136
Plato's cosmology 132
politics of architecture 141–142
Popper, K. 135–136, 141, 167

Index

Post-growth (Jackson) 28, 30
post-occupancy evaluation (POE) 112, 265–166
poverty 2, 6, 12, 42, 159
pre-1919 houses 41–42
problem(s): airtightness 89–94; determining outcomes 44–57; insulation 62–68; occupants 111–120; overheating 70–74; renewables 97–106; ventilation 78–84
Pro Clima membrane 63
The Progress of This Storm, Nature and Society (Malm) 20, 36–37
Project Calebre 54–55, 69, 92–93, 116
Prometheus programme 72
pseudo-Modernism 2, 6, 15, 20, 62, 144, 161, 167
psychic automatism 149
psychrometric chart 63
Publicly Available Specification (PAS 2030) 97, 108, 109
Publicly Available Specification (PAS 2031) 109
Publicly Available Specification (PAS 2035) 44–47, 78, 81, 108–109, 124, 127, 164–165

Q-bot 65

radical politics, and Modernism 143
rainwater and greywater harvesting 100
Randers, J. 7
rebound effect 112, 115
Reduced data Standard Assessment Procedure (RdSAP) 97
Refurbishing the Nation – Gathering Evidence 89
Regeneration and Retrofit (UKGBC) 109
Relish Smartwire 118
Relish™ 117–118
renewable energy 7
renewables 97–106
RE:NEW Technical Risk Matrix 108
'rent-a-roof' scheme 105
retrofit: architecture 37–38; energy case for 43–44; future of 164–168; leaders, architects as 126; origins 61; and performance 124–126; pre-requisites for 60–68; progress of 123–127; realigned 133; representation of society as a work of art 160–161; as subversive art 162; at urban scale 60–61; utopia and 134–137; Webb's lessons for 156
Retrofit Consumer Charter 108

Retrofit Coordinator 108, 164–165; role of 166
Retrofit for the Future Competition 44–57, 86–87, 107–120, 164
Retrofit for the Future sites (IfS) 115
RetrofitWorks 25
Ridoret of La Rochelle 83
'right to buy' scheme 17
Ritchie, H. 13
Roark, H. 19
Roberts, D. 49
Rockwool insulation 65
romantic pessimism 22
rooftop photovoltaics 103–104
Rousseau, J-J. 132
Rowe, C. 137, 149
Royal Festival Hall 1

Sapiens (Harari) 8
Scaling up Retrofit 2050 (IET 2018) 50–51
Schlemmer, O. 147
Schulz, N. 138
Schumpeter, J. 142
scenario planning 139
self-regulation 20–21
Self, W. 140
Sennett, R. 140
Share Power: How ordinary people can change the way capitalism works – and make money too (Webb) 6
Sinclair, I. 145
Situationism 137, 150
Skidmore Report 6–7
Smart Export Guarantee (SEG) 102
Social Housing Decarbonisation Fund 56, 123–124
social housing retrofit schemes 104, 124
social media 38
social mobility 6
society as a work of art 159–160
solar industry 102
solar panels, Australia 103
solar thermal 9, 70
solid masonry walls, 63, 66, 70, 77, 87, 116
Southway Housing Trust in Manchester 118
Standard Assessment Procedure (SAP) 77, 97, 115; ratings 41–42, 70
starchitects 125
Stoknes, P. E. 7, 139
STO-solar insulation 65
Surrealism 149–150
sustainability 13, 36
Sustainable architecture, notion of 8

sustainable development: goals 7–8; origins and implications 13
Sustainable Energy's PlanLoCaL 105–106
sustainable retreat 26–33; non-growth society and 28; technology and 27
Sustainable Traditional Buildings Alliance 43, 108
Swansea Communal Solar 104

technology 27, 49
Technology Strategy Board 81
thermal comfort 73
Thermal Delight in Architecture (Heschong) 84
This Changes Everything, Capitalism vs. the Climate (Klein) 141–142
Thompson, R. 88
3D virtual model 37
Thunberg, G. 134
Today's Attitudes to Low and Zero Carbon Homes (NHBC 2009) 80
Tomorrow's Economy (Stoknes) 7
Towards the delivery of a national residential energy efficiency programme (Arup) 109
TrustMark 108–109
turbo-capitalism 28–29
2020–21 English Housing Survey 41

UK Centre for Moisture in Buildings 108
UK Green Building Council 109
UK Net Zero Carbon Buildings Standard 53–54
UK's: Carbon Budget 41; energy 41; 'fit and forget' tradition 92; retrofit programme 127
'*The Uncanny*' (Freud) 145
unemployment 139
United Kingdom: appeal of PV panels 100; 'fit and forget' culture in 81; PVs installation in 101; social housing retrofit schemes 124; ventilation levels in new homes (2019) 90; waste water heat recovery 100

United States: poor ventilation in 45
University of Strathclyde (ESRU 2005) 88
urban ecology 151–157, *153*, *155*
utopian ideals of modern architecture 139
utopianism 139–140
utopian mission 165–168
utopia, retrofit and 134–137

ventilation 45–47
Ventive 82
Venturi, R. 2, 137
Vitruvian Triad 14

Wainwright, J. 7–8
Walter Segal tradition 49
Warmer Homes, Greener Homes: A Strategy for Household Energy Management from 2010 (GOV.UK) 69
waste water heat recovery (WWHR) 100
Webb, M. S. 6
Webb, P. 154–156
Welsh 'Mini Power Stations' project 104
Which magazine 100
WHISCERS 66
White, C. 44, 46, 60
White Design 46
WHO 90
whole dwelling approach 108
Wilson, H. 1
Woking's Combined Heat and Power plant 104
work/life balance 28
World War I 1, 27
World War II 27
WUFI 63

ZED Factory 46
zero carbon 27, 51–52, 53–55; construction 8; emission 37; technology 70
Zero Carbon Hub 88

For Product Safety Concerns and Information please contact our EU representative GPSR@taylorandfrancis.com Taylor & Francis Verlag GmbH, Kaufingerstraße 24, 80331 München, Germany

Printed and bound by CPI Group (UK) Ltd, Croydon, CR0 4YY
10/01/2025
01818724-0001